The Origin of Scientific Revolutions: The Clash Between Pervasive Research Traditions

Rinat M. Nugayev

The Origin of Scientific Revolutions: The Clash Between Pervasive Research Traditions

Rinat M. Nugayev

Library of Congress Cataloging-in-Publication Data
Names: Nugayev, Rinat (author)
Title: The origin of scientific revolutions : the clash between pervasive research traditions | Rinat Nugayev
Description: Washington : Academica Press, 2024. | Includes references.
Identifiers: LCCN 2024936208 | ISBN 9781680533071 (hardcover) | 9781680533088 (e-book)

Contents

INTRODUCTION.

Subtle Mature Theory-Change Models and Relentless History of Science Data.

> *"If history is to rise above a mere collection of anecdotes, it must be written from some point of view and with some unifying theme. It is here that the philosopher has a contribution to make by supplying some distinctive perspective, such as Kuhn's view about paradigms, normal science and revolutions."*
> — **William Shea**

It is taken for granted that the American philosopher, historian, and sociologist of science Thomas Samuel Kuhn, along with his partisans, stood at the origins of the so-called '*Historical Turn*' in the airy-fairy philosophy of science realm. Therein the assiduous researchers increasingly commenced to realize that '*to comprehend what science is, one needs to* comprehend how it advances.' The exponential growth of history-of-science trailblazing rational reconstructions, obliged to the breakthrough, overshadowed for the first time distinctive epistemological ideas that stood behind the path-breaking turn. It engendered the enticing impression that the dazzling writings of the epoch of "*die Sturm und Drang*" and, above all, the stupendous "*Structure of Scientific Revolutions*," first published in 1962, constituted a *downright* generalization of history-of-science impelling data. Whereupon there transpired bewitching illusion that to do away with logical empiricism, it was enough to just punctiliously 'wipe your eyes' and behold with a new, unbiased, fresh gaze at bothering history of science. And novel epistemological discoveries will inexorably spill out of abundance.

Though sober elucidation of what significant epistemological results were *really* obtained supplanted precocious enthusiasm. This, in particular, was convincingly evidenced by the collection of Thomas

Samuel Kuhn's latest writings "*After the Structure of Scientific Revolutions*," already published after his death.

> "Now I think *we overemphasized the empirical aspect of our enterprise* (an evolutionary epistemology need not be a naturalized one)" (Kuhn 2000, p. 95; my italics).

It is no wonder that in the collection one of Kuhn's ultimate goals was not only to indicate the basic changes in his epistemological position that gradually took place after 1961 but also to manifest their *philosophical premises*.

Accordingly, the ardent appeal to the epistemology of Immanuel Kant ('Ding an sich'), Hilary Putnam ('internal worlds'), to Martin Heidegger's inextricable philosophy ('being-in-the-world') is frequent and apparent. Yet the blunt attempt relentlessly elicited the dark side of the coin.

Ever since "*After the Structure of Scientific Revolutions*," it becomes more and more clear that even the amplest, detailed, elaborated, and the sophisticated doctrine of modern history & philosophy of science constitutes such a *hotchpotch* of diverse philosophical/epistemological/ sociological/psychological/historical views, biases, values, and opinions that it makes its objective assessment and active application in common research extremely difficult.

Alas, this is not accidental and in no way is Thomas Kuhn's or somebody else's fault. The puzzling situation is a downward reflection of the 'hard fact' that real, 'alive' resolute scientific revolution is a jolly intricate and inextricable mixture of the most varied (and often rather unexpected) factors, ranging from the social-psychological features of scientific communities up to the odd social -cultural 'atmosphere' in which the members of the communities have to ply. (Not to forget sometimes dangerous political and economical peculiarities like the second world war or French revolution). Perchance, it will take generations of philosophers, sociologists, and historians of science to salve the corresponding problems. In my humble part, I dare to indicate a *promising direction* of future research activity at best.

Discreet *descriptive* methodology vividly embodies the hopelessness of searching for an effective theory-choice resolution version. Its advocates gingerly reckon that the methodologist's professional business is to describe punctiliously the subtle criteria *actually* employed by scientists. They regrettably maintain that cultivating methods, norms and sublime ideal cognitive models is a hopeless task.

For instance, the lucid falsificationist account of scientific history as a process of proposal and subsequent refutation of different hypotheses is unacceptable for a descriptivist not because it often contradicts historical data, but since it is a confounded *theoretical* description. A historical juncture is depicted from the professional standpoint of a historian or an epistemologist, but *not* from a *real participant* at the events.

In my view, the descriptivist sober arguments against the normative methodology (see Maurice Finocchiaro's multifarious writings, for instance) are grounded on setting the logical-theoretical descriptions in sublime philosophical minds *against* the logical-historical descriptions in the mundane minds of historical agents. The descriptivist arguments would be immaculately correct if these descriptions were equally *complementary*, tantamount views just held by different scholars on peculiar historical events.

However, in my view these types of descriptions refer to different *levels of comprehension*. The descriptivist picture refers to an *empirical* level, and the normative standpoint - to a more profound *theoretical* one. The two levels are so *inextricably* linked that sophisticated theoretical description relentlessly 'shows itself' through the descriptive one. Vague and subtle, so to speak, whimsical 'laws' of scientific methodology abode only as the aspects, the *facets* of individual actions of certain scientists.

Let me scrutinize subtle interrelation of theoretical and empirical inextricable levels of methodological studies. To give a sense of the interrelation, I would prefer to contemplate an oversimplified example from statistical thermodynamics - just to catch the congenial hallmark that is perchance common for all the models both in natural science and in humanities as well.

After Immanuel Kant, it is a platitude that a simple description of a period of scientific development can buttress to ascertain an *empirical* regularity at best. To elevate it to the ranks of a genuine theoretical proposition, i.e. to bestow on it gorgeous status of *universality*, we are bound to convert the empirical regularity to an expression stiffly connecting certain terms of *theoretical language.* To be more exact, let me scrutinize a guileless example borrowed from Vyacheslav Stepin (2005): the renowned Boyle' law.

Behold the Boyle fine equation derived on the basis of some sedulous observations. (On the metahistorical level this corresponds to a certain apparent regularity ascertained for a certain amount of theory-choice processes). One has carefully examined N gases (sedulously have inquired into the transitions 'Ampére-Maxwell,' 'Lorentz-Einstein,' and 'Planck-Bohr,' for instance). Nevertheless, no increase in the number N of gases considered (additional survey of the 'Bohr -Schrodinger' transition) can ever elevate the discovered empirical regularity up to universal law. *No* finite number of observations can guarantee against the regularity being broken in some future observation or experiment. Moreover, such a violation is inevitable in the distinctive case of gases under high pressures at least. This case is exactly displayed by the complicated Van der Waals equation.

To proffer the fine theoretical law describing the evident connection between pressure and volume of a gas, the immaculate *ideal model* must first be punctiliously constructed. The set of abstract objects making up this model are ideally rigid and infinitesimally small particles, the so-called 'material points' that untiringly collide with each other. These bogus particles move according to the strict laws of classical mechanics, vigorously striking the walls of an experimental vessel. As a result, the pressure of the gas is properly defined as the sum force of their blows per unit surface area. An immaculate mathematical inquiry into the relations between these abstract objects makes it possible to *theoretically reproduce* the renowned law PV = const previously established only experimentally.

Consequently, the inductively obtained abject expression becomes an immaculate theoretical law, correctly describing the behaviour of sufficiently rare gases. The decrepit empirical equation PV= const is now digested into a sublime theoretical law. Whilst the form of the equation remains the same, P and V epitomize the fine relations between abstract objects of a theoretical language now, rather than the dubious relations between real empirical gases and their vessels.

So, since the predictive force of inductive generalizations is substantially probabilistic, the extrapolation of an empirical regularity is not enough to set it above the ranks of hypothetical supposition. The required transition is possible only when the connection between the quantities of blunt empirical expression is reproduced as a result of operations on the abstract ideal objects that make up *the ideal model* of the particular domain of reality.

To provide the genuine law describing the connection between volume V and pressure of a gas P, the ideal gas model should be punctiliously constructed and the empirical regularity properly rederived from the ideal model.

The abovementioned is in no way to buttress the naive realist stale position asserting that the theoretical law obtained is a more accurate 'picture of reality.' On the contrary, as Nancy Cartwright (1983) shrewdly put it, mother Nature on her own may be lacking profound and uniform regularities. Most of the regularities one possesses may represent only the subtle ways in which we contrive our theories (and construct the corresponding experimental devices) to brood on things. As for ideal models, Cartwright retorts that in several branches of quantum mechanics one easily and cheerfully uses in bold 'Ptolemaic fashion' a whole bundle of models of the same phenomena. Incidentally, nobody holds that a model of the throng constitutes the 'whole truth.' Moreover, the models may be even mutually *inconsistent* in true 'Ptolemaic' fashion. Ever since the models constitute primarily the intellectual tools that help one to 'save the phenomena' and build the corresponding pieces of experimental technology. They enable one to intervene effectively into the processes and to create new and hitherto unimagined phenomena (recall Ian Hacking's fine 1987 book).

Furthermore, our common idea of approximation consists in that, to avoid a mess, one writes down an equation that is true only approximately. But although some approximations are leading *away* from the truth, there are far more approximations *towards* the truth. Phenomenological laws, arranged as a result of approximating the fundamental laws to concrete situations, are *closer to the truth* than the fundamental ones. For obvious reasons, most of the fundamental laws describe extremely oversimplified and unilaterally idealized situations that can have nothing to do with actual experimental facts (see new-fangled superstring theory).

For instance, the well-known four-potential of the electromagnetic field in Maxwellian electrodynamics $A^i = (\mathbf{A}, \varphi)$ has no *direct* empirical meaning serving only as a powerful means for the mathematical description of electromagnetic phenomena.

Though if the basic idealized models of a theory can be jolly far from the truth, *what are they for?*

- Firstly, they supply one with useful theoretical *vocabulary* to describe the empirical data and to comprehend quite novel phenomena. Moreover, most of the components of an ideal model are sheer *idealizations*, i.e. the abstract objects that represent *caricatures* of real objects. Queer caricatures possess only part of the properties of real objects. For instance, a material point of Newtonian mechanics shares only one property with a common bullet – 'to have a mass.' Yet one resolutely inserts the material point into the ideal gas model and then applies the laws of classical mechanics to its components striving to exhibit that the property 'to have a mass' is necessary and sufficient for a model to reproduce a certain realm of mechanical events.
- Secondly, especially in the most advanced natural sciences, they help to *predict* new phenomena and to construct experimental devices to check the predictions.
- Thirdly, if the idealized model cannot proffer a proper explanation, it continues to embody a "system of mathematical propositions, deduced from a small number of principles, which aim to represent as simply, as completely, and as exactly as possible a set of experimental laws" (Duhem [1906], 1954, p.19).
- And the last, but not the least, advantage of having an ideal model consists in that the model can delineate *the domain of validity* of our solid knowledge more exactly. For instance, the basic premise of Boyle's fine ideal model consists in that one takes the gas molecules as material points that can interact during the collisions only. The premise indicates that Boyle's model can exhibit the behaviour of rarefied gases only.

Likewise, the recipe for salving the theory-change situation in history of science inductively obtained at the expense of examining the *already resolved* theory-change situations can only be taken as an empirical regularity at best. To be called *universal,* i.e. to attain the status of genuine theoretical proposition, it must be tentatively derived from the epistemic *ideal model of the theory-change process.* This model should *theoretically reproduce* the theory-change process by examining the subtle relations between the following fine structures.

(1) The *cognitive subject* (the scientific community in general or an influential group of researchers);

(2) the necessary internal and external *conditions* and *factors* of its creativity;

(3) the *results* of its cognitive activities – vague ideas, brand novel hypotheses, powerful and solid theories, etc.

The relations between the theoretical objects of the epistemic (gnoseological) model should be depicted by the abstract philosophical principles borrowed from multifarious current epistemological doctrines. On the other hand, an ideal model is intended to treat *its own facet* of the inextricable mellow theory-change process, consistently and persistently ignoring those peculiarities of the scientific revolution that it regards as contingent ('inessential').

The specific nature of an epistemic model should be determined by its system of basic abstract objects, which in turn are grounded on those empirically fixed features of the cognition process that necessarily lead to the theory-change. The fundamental ideal theory-change model can be as far from truth as Newton's fundamental theoretical scheme of classical mechanics is. The latter describes the relations between such ghosts (ideal objects) as the 'material points,' 'forces,' and 'inertial reference systems.' Wherein a *partial theoretical model* properly constructed from the basic epistemic one to scrutinize the concrete theory-change situation may be closer to the truth, to what 'really happened' in the history of science.

Punctiliously taking into account the peculiarities of the abovementioned theory-change process, one can construct such a sequence of ideal models that are closer *to the empirical level,* i.e. to what had been *perceived* by real participants at the events.

Moreover, to reproduce the inextricable interaction of external and internal factors in a scientific revolution theoretically, one has to examine first the workings of each factor independently of the others, in so to say '*purified form*.' And only after that, he obtains an opportunity to construct a model tentatively describing the inextricable interaction of the sociocultural factors with the internal or 'cognitive' ones.

The account of the accomplished theory-change process, grounded on the ideal epistemic model, will inevitably constitute the history of events that were chosen and interpreted in a *normative* way. Such an inquiry will constitute an unavoidable *reconstruction* of real history since it will embrace only those facts of scientific history that are necessary for the model. On being applied to the past, the ideal model will ascertain the *rational reconstruction* of real theory-change process. Being applied to the present state of the development of science, it will consolidate some problem-solving *standards*.

The subtle relations between subtle theoretical models and hard empirical facts can be illustrated by the following example (Norton Wise 2017, pp.347-348).

S. Gravner and D. Griffeath (2009) have contrived a simple computational model, which grows a virtual snowflake from a small seed of ice surrounded by water vapor and governed by three mechanisms merely: (1) diffusion of water vapor from the crystal; (2) freezing and melting in a narrow boundary layer; (3) and attachment rates at the boundary that favor concavities.

Eventually, Gravner and Griffeath forthrightly acknowledge that it is not very clear just how their intuitively plausible parameters correlate with physical processes. Moreover, their simulations do not treat important issues of non-symmetry, randomness, singularities and instabilities. Nevertheless, they believe that the evolutionary simulations provide genuine *explanations* of many of the characteristics of real snowflakes, "both in general morphology and in the details of their traits."

All in all,

> "the simulations generate an evolutionary narrative which explains the natural order of snowflakes as an essentially historical order" (Norton Wise 2017, p. 354).

The budding analogy between ideal models in natural science and humanities leads to the interpretation of theory-change models that can better be comprehended in terms of Max Weber's congenial notions of *action, social action*, and historical and sociological *ideal types.*

The concept of *ideal type* is pivotal for Weber's solid methodology of historical and sociological cognition. For the celebrated author of "*Die Protestantische Ethik*" an Ideal Type cannot be directly derived from empirical reality but is sedulously constructed as a theoretical scheme. Weber wittily called the ideal types 'utopias.'

> "The more sharply and unambiguously the ideal types are constructed, the more they are aligned to this world, and the better they fulfil their task, - in terminological, classifying, and in heuristic respects as well" (Weber 1964, band 2, p.15).

Weber's ideal types are apparently analogous to ideal models of natural science. Ideal types "may be as rarely met in reality as physical reactions that are calculated under the assumption of absolutely empty space" (Weber, 1964, band 2, p.10). Such substantial notions as 'economical exchange,' 'homo oekonomikus,' 'craft,' 'capitalism,' 'Christianity' are typically ideal constructs, serving as a powerful *means* of describing real individual historical formations. Incidentally, one must make out the *sociological* and *historical* ideal types. Both have a "character of some utopia, emerging under conceptual amplification, isolation of certain elements of reality" (Weber 1964, band 2, p.190). But there are substantial differences though.

> "Sociology, as has been presupposed many times, contrives the notions of types and looks for *general rules* of events, unlike history, which seeks to give a causal analysis [...] of *individual,* culturally significant actions, formations, persons" (Weber, 1964, p.545, band 2; my italics).

In sociology, ideal types are more universal than in history; they are called 'pure ideal types.' (For instance, a sociologist strives to construct ideal models of domination that can be found anywhere at any time). Sociological ideal types are characterized by the following hallmarks.

(i) The human actions which they encompass are constructed as if they have a place in ideal conditions.

(ii) The sociological ideal form is independent of local conditions of place and time. Hence, while a historical ideal type serves to isolate the causal relationships that occur at least *once*, sociological ideal types isolate the relationships that exist *constantly*.

(iii)Sociology for Weber is a peculiar discipline that scrutinizes *social actions*.

> "Action is the name given to human behaviour regardless of whether it is an external or internal act, inaction or suffering, if, when, and in as much as those who act give it a *subjective* meaning. But 'social action' should be the name given to such an action which in meaning, - as implied by those who act, - is related to the behaviour of others, which in turn maps out its course" (Weber, 1964, band 1, p.1; my italics).

One has to acknowledge the deep connection of Weberian methodology with Heideggerian ontology stating the central role of Dasein as a pivotal subject of cognition. Heidegger never tied of emphasizing that Dasein is not a thing, substance or merely object. Rather, it is an accumulation of end-directed, or intentional, *actions*:

> "The person is no thinglike and substantial Being [...] Essentially the person exists only in the performance of intentional acts, and is therefore essentially not an object" (Heidegger 1927).

The abovementioned connection between Weber and Heidegger will play a decisive role in working out the methodology of mathematical projections of nature outlined in Chapters I and II of this monograph. The methodology is grounded on the assertion that 'mathematical projection' constitutes an existential phenomenon that serves as a condition of possibility for *both t*heory and practice in the sciences.

'Social action' is a genuine '*atom*' of Weberian keen sociology and cannot be divided into pieces. Moreover, neither whole society nor smaller forms of collective organizations can be taken as authentic subjects of action. Only individuals can. Thus, the application of such vague and controversial notions as 'nation,' 'state,' etc. is not strictly forbidden in Weberian sociology. Yet Weber insisted that one should never forget that they are *not* genuine subjects of social actions.

The author of "*Die Protestantische Ethik*" famously elucidates four types of social action. The two of them are *irrational*, in the sense that they are not the results of deliberate choices. These are *traditional action*, guided by habit, and *affectual action*, proffered by emotions. The further two are types of rational action. The first is 'Wertrational Aktion' that is undertaken as a result of belief in the ultimate value of acting in a certain way. The second is 'Zweckrational Aktion' that constitutes a calculation of the appropriate action to be taken to achieve the desired end. Weber contended that in modernizing society zweckrational action, precisely because it is calculable and predictive, becomes the dominating type of social action.

This action is the most comprehensible one since it:

(1) is aimed at achieving those goals that an individual is aware;

(2) employs for this attainment the means that are taken quite adequate by the individual himself.

Note that Weber does not consider zweckrational action as universal or even of particular significance in empirical reality. On the contrary, it is an *ideal type* of action, a *paradigm* of social action, just a scale for comparison with other types of action. Zweckrational action is a cognitive rather than ontological notion. It is a helpful *means* of siffing reality but is not reality itself. Herewith true behavior of an individual is commonly directed simultaneously towards *several* types of action. It contains them as its inherent facets.

Let me recall the theory-change epistemic model now. Thanks to Weber's vocabulary, this model constitutes the ideal sociological type of scientific action. This sociological type stands in resolute opposition to historical types used by descriptivists. The epistemic model as a paragon of all the ideal types reconstructs scientists' social actions as if they have a place in ideal conditions. Besides, the ideal norms it fixes are independent of local conditions of place and time.

Howbeit, an ideal epistemic model can serve as a useful *scale* for measuring the inevitable deviations from ideal norms and as a useful *paradigm* for disquisition of new theory-change situations taking place in science. More importantly, one must make up *two kinds* of ideal epistemic models - the Basic Ideal Model that is jolly abstract and is extremely far from reality and the Particular Ideal Models constructed from the Basic one for examining the concrete theory-choice situations. (Just as Weber masterfully had extracted his ideal type 'Capitalism' from the concrete analysis given in his ground-breaking '*Die Protestantische Ethik*'). Note that an epistemic model describes first and foremost the aspects of an *individual* scientist's activities; the activities of the scientific community as a whole or even of its influential parts are described only indirectly.

However, on introducing an ideal epistemic model that does not set up from history-of-science data direct generalizations, one inevitably comes up against the following problem. Let one assume that he has two (or more) different basic ideal models, i.e. two (or more) ideal types of theory-change, which can be applied to a theory-change process. As I have already pointed out, to choose between two basic models, one has to compare them as the tools for scrutinizing concrete theory-choice situations. One must compare them as efficacious *instruments of comprehending*. Therefore, the definite theory-change period should be chosen, and two particular reconstructions of the period produced by diverse basic models should be advanced and compared. Yet which reconstruction should be chosen? Appealing to 'hard facts' can turn out pointless since the facts, as a rule, are fabricated by the models. Different models commonly elicit different facts.

For instance, falsificationists, following Karl Popper, will stress and multiply examples of theory refutation, such as Maxwell's refutation of the Weber-Neumann theory, Einstein's refutation of Newton's theory, etc. Inductivists, following Francis Bacon, will appeal to inductive generalizations such as Kepler's startling laws, Faraday's renowned laws of induction, and so on. Even when facts are shared, they are weighted differently from one model to another.

Whereupon, how can notorious disparity between the models and the facts be detected when the facts are chosen by the models? In other words, what are the *meta-criteria* of choice?

Applying Max Weber's sober evaluation of scientific action as the closest to zweckrational one and his pivotal notion of the social ideal type, one can conclude the following. Viz., that particular ideal model should be chosen, according to which most history-of-science data can be understood from a *unified* stance, and most contemporary scientists' expert opinions about the given period appear to be correct (chapter II). Moreover, the ability of the model not only to comprehend the established history-of-science facts but also to help to discover the new thrilling ones is also significant.

Nevertheless, one should not ignore the fact that the notorious 'confrontation model' (Lakatos, Laudan) of the relationship between history & philosophy of science, according to which history of science is taken as a 'repository of facts for testing theories of scientific change' (Arabatzis 2017), is now considered obsolete. Rather, there are grave reasons that historical facts can be sufficiently detached from epistemological models so as to be applied in their evaluation (Nickles 1986). Thus, it has been soberly suggested that philosophy of science and epistemology should be considered rather as a '*hermeneutic enterprise*' that merely interprets the historical record in terms of its analytic concepts. And the latter in the process of interpretation may be refined or modified.

Yet the proper application of Weberian methodology is tightly connected with the important circumstance.

Forsooth, the sober answers (grounded on the well-known epistemological models) are reduced to the following ones.

(I) inductivism (Whewell); (II) falsificationism (Popper); (III) conventionalism (Duhem); (IV) sophisticated falsificationism (Lakatos);(V) social-psychological version (Kuhn).

Some substantial drawbacks of the (I- IV) options are determined by one and the same reason – lack of scrutiny of relations between cognitive and social aspects of science. The epistemological models ignore (to varying degrees) the hard fact that science is a substantially '*community effect*.' They hold a 'great genius' vision of the progress of science: the startling insights of peculiar individuals, the 'chosen ones of the gods,' are what drive science forward. However, this is defiantly *unilateral* picture. The luminaries (Newton, Einstein) can *start* a scientific revolution – but only when a broader community of researchers properly prepared the field.

Hence sociology of scientific knowledge (SSK) in the wake of (V) decidedly broke away from positivist view of advancement of science as an asocial process revealing universally valid truths. Overcoming the limitations of Kuhn's reduction of sociology to social psychology, SSK took scientific knowledge as an *artificial* product of social, political, and economic interactions.

However, the dazzling picture fatally leads to '*incommensurability* tenet': the partisans of 'old' and 'young' paradigms live in inscrutably alien socio-cultural worlds divided by scientific revolution. Yet since the 'old' and 'new' paradigms are radically separated from one another by impenetrable 'incommensurability' barrier, it is baffling to see how one gets from one to the other. Hence Kuhn uses quasi-religious metaphor of 'conversion to belief' instead of properly explaining *how exactly* new ideas emerge against the background of their predecessors, from which they are so totally distinct.

Logical empiricism and SSK are opposite sides of shabby coin. Though we know, partly from Kuhn and Collins, that *tensions* between the scientific practices are *crucial motor* for change.

The *aim* of the present monograph is to propose such an *ideal model* that harmoniously reconciles cognitive and social aspects of the development of science and provides analytical tools for studying the social mechanisms by which the cognitive structures of scientific knowledge evolve and interact with each other. And to exhibit how the model 'works' reconstructing the paradigm case studies (Ptolemy-Copernicus breakthrough; Maxwell's synthesis of optics & electromagnetism; Planck-Einstein upheaval, general relativity contrivance, Dirac's unification of special relativity and quantum mechanics).

The author's aim is to strengthen efforts in landing the sky-high epistemological models of scientific revolutions. In the wake of Kuhn's '*Structure*' amended by Foucault's 'practice' and Latour's sociology of knowledge, as well as the writings of Stephen Shapin, Philip Kitcher and Helen Longino, a further step is taken on the path of reasonable explanation for the mundane reasons for the triumph of the 'novel' paradigm over the 'old' one. It is contended that a corresponding expansion of the epistemological base of research requires the involvement of phenomenological ideas presented by Edmund Husserl and Martin Heidegger. They consider the history of human cognition as a series of epochs of 'unconcealment' with the integrity of each one provided by reconciliation, interaction and interpenetration of sundry practices.

Profound breakthroughs in science were first and foremost *not* due to ingenious contrivances of brave novel paradigms or invention of 'send by God' immaculate ideas 'ex nihilo.' On the contrary, the breakthroughs were caused, among other things, by the harrowing piecemeal processes of accommodation, interpenetration, and intertwinement of the 'old' research traditions preceding the radical breaks.

Creation of a new mature theory is by no means the result of a sudden individual insight (or uncanny 'Gestaltswitch'), but the mundane payoff of *piecemeal* reconciliation of several research practices.

It is via 'language games' of multifarious conceptual resources that a new math is born, destined to form the core of prolific research tradition. This suggests a conservative though flexible mode of a scientific upheaval: it is *not* substantially different from other types of socio-cultural changes.

In good sooth, the previous epistemological accounts have intensely oscillated between *two alternative extremes*. On the one hand, in the taken for granted facile vein, the apparent differences between research traditions were hastily taken to be insignificant and fruitful communication (and even deep interpenetration) unproblematic.

On the other hand, in the new-fangled, post-Kuhnian distinctive disquisitions, really significant differences between the research traditions are commonly posited to be so *radical* that their actual communication is regarded almost impossible.

The present study stems from a more mundane, 'unhagiographic' *intermediate picture*. Respectively, I readily admit that the substantial differences between the research traditions really exist at the various levels, ranging from entrenched ontological commitments and inextricable epistemological beliefs similar to Ptolemy's notorious instrumentalism up to subtle mathematical and experimental technique.

However, these often antagonistic traditions were able to *communicate* fruitfully in the creative acts of such pundits as Ptolemy, Copernicus, Galileo, Newton, Maxwell, Einstein. Owing to the *intersubjective* socio-cultural nature of human knowledge (Longino), the research practices had vigorously communicated by the various ways that permitted comparisons, adaptations, and even fruitful cross-fertilizations.

Respectively, **Chapter I** is devoted to identifying the system of social and cognitive prerequisites of reconstructing a scientific revolution grounded on the 'paradigm of paradigms' – the Copernican revolution. **Chapter II** highlights the theory-change model itself, while **Chapters III, IV, V,** and **Conclusions** deal with exhibiting the opportunities of the model proposed in grappling with the obtrusive philosophy of science problems via manifold case-studies. The latter embrace Maxwell's synthesis of optics & electromagnetism; the Planck-Einstein upheaval, the general relativity creation, Dirac's unification of special relativity and quantum mechanics and even Hawking's amalgamation of general relativity and quantum field theory. Wherein, while the **third** chapter deals with the incommensurability problem, the f**ourth** one is dedicated to the cross-contradiction eliciting. In continuation of this the **fifth** chapter reveals the role of metaphysics.

Viz., the sober origins of a scientific revolution are revealed as a result of the following 'case studies.

(a) **The Copernican Revolution** is elicited in the distinctive context of the intense clash, subsequent interaction, and profound interpenetration of Aristotelean and Ptolemaic sophisticated theoretical languages. Aristotelian – Ptolemaic heathen cosmology could not help but be exposed to repeated severe Christian and Moslem attacks during the Middle Ages since it confronted the stout principles of monotheism not admitting the impenetrable, strict demarcation line between the celestial and mundane realms. All seemingly different worlds should have the same Creator. Henceforth the Copernican relentless breakthrough should be evaluated in the refined scope of further clarification of the tremendous gap between astronomy and physics, as well as the subsequent commencement of effective efforts to eradicate it in the favorable monotheistic social-cultural context of Christian Weltanschauung enforcement. Copernicus – probably against his will - paved the way for Galileo's and Newton's mighty mathematical physics, and namely, that constituted the grave reason for his victory over Ptolemy. If the robust demarcation line between divine and mundane worlds is lacking since the Earth is just an ordinary planet of the Solar system, the mathematical subtle notions, and elevated principles should apply both to its rotations around its axis and the Sun, as well as to all the bodies moving along its surface. The posterior indispensable contributions of Galileo Galilei, Johann Kepler, Rene Descartes, and finally Isaac Newton incarnated the stiff milestones of the mathematics descent from sublime Skies to sinful Earth and the reciprocal ascent of stout Earth physics in comprehending the refined and distinctive Divine phenomena.

(b) A case study of the genesis and functioning of a neutral mathematical language related to the **Maxwellian synthesis** of optics and electromagnetism (contra the incommensurability tenet) is provided. It is elicited that its basis is constituted by the mathematical language of continuum mechanics. The main function of the neutral language was to project the consequences of all the unified theories onto the mathematical model, 'rewrite' all known laws in this mathematical language, compare their conclusions with each other to eliminate contradictions and generalize the stuff in a self-consistent system of equations.

(c) It is contended that the **Quantum and the Relativistic scientific revolutions** were contemporaneous since they had the *common origin* – a tough skirmish between the profound classical research traditions. (The traditions of Newtonian mechanics, Maxwellian electrodynamics, Boltzmann's statistical mechanics and Carnot and Clausius's thermodynamics). The breakthrough was embodied in Max Planck's persistent attempts to comprehend electromagnetic phenomena through the lenses of statistical thermodynamics. Planck was clearly aware of the thorough cross-contradiction between 'the physics of material bodies' and 'the physics of the ether.' Moreover, he outlined the chary sketch of its attenuation and even withdrawal: the paradigms 'must be modified to remain compatible.' And Planck himself took the first promising step in modifying the physics of the ether: 'not only matter itself but also the effects radiated from matter possess discontinuous properties.' Einstein's substantial part consisted in that he took the next step in modifying the second component of the encounter – the physics of material bodies. Einstein's risky light quanta hypothesis and discreet special theory of relativity turn out to be mere milestones of the unwinding of Maxwellian electrodynamics and statistical thermodynamics reconcilement research program.

d) One of the turning points of the history of XXth century physics – Paul Dirac's stupendous contrivance of the relativistic theory of the electron – is elucidated in the context of apt theory- unification model posited in the preceding chapters. The rational reconstruction of the **Dirac synthesis** of special relativity and quantum mechanics is provided that makes it possible to compare it with the classical synthetic samples (Newton, Maxwell, Einstein) and to unfold its specificity. The fine structure of Dirac's intricate research program and the odd principles that put up its powerful heuristics is scrutinized with special emphasis on the highly controversial tenets of 'mathematical beauty' and 'unity of science.' Digging still deeper into the background, it is contended that despite the relentless Dirac's remarks denigrating the role of epistemology one can trace a tinge of its indirect influence epitomized in Husserlian quasi-Platonic doctrine through Arthur Eddington's and Hermann Weyl's whimsical mathematical writings embracing the pioneering efforts to merge gravitation with electromagnetism. Accordingly, the milestones of Dirac's research program realization are elicited.

CHAPTER I.

Scientific Revolution as a Substantially Social Process: Powerful Lessons From Copernican 'Paradigmatic Case of All Paradigms.'

"*You have brought the internal and external histories of science, which in the past have been very far apart, together*"
— Alexander Koyrè's remark in his post-Structure letter to Thomas Kuhn

"*Their [the Ptolemy partisans'] experience was just like someone taking from various places hands, feet, a head, and other pieces, very well depicted it may be, but not for the representation of a single person; since these fragments would not belong to one another at all, a monster rather than a man would be put together from them*"
— Copernicus. 'De Revolutionibus'

Abstract.

Since Alexander Koyrè and Thomas Kuhn, the Ptolemy-Copernicus breakthrough is taken as the 'paradigmatic case of all paradigms,' and many weighty history &philosophy of science problems and their solutions are commonly contemplated precisely on this material. In particular, at least some of the drawbacks of well-known epistemological models of the Copernican revolution turn out to be connected with one and the same reason – lack of proper scrutiny of the subtle interconnections of cognitive and social aspects of scientific research. The epistemological models, as a matter of fact, ignore to varying degrees the hard fact that science is a 'community effect.' According to the romantic 'great genius' standpoint, the startling insights of very peculiar individuals, the 'chosen ones of the gods,' are what drive science forward.

Though this is an apparently distorted picture and the goal of the first chapter is to strengthen efforts in landing the sky-high epistemological models of scientific revolutions. In the wake of Kuhn's 'Structure' amended by Michel Foucault's 'practice' and Bruno Latour's sociology of knowledge, as well as the writings of Stephen Shapin, Philip Kitcher and Helen Longino, a further step is taken on the path of reasonable explanation for the mundane reasons for the incipience and triumph of the Copernican breakthrough program. It is contended that a corresponding expansion of the epistemological base of research requires the involvement of phenomenological ideas presented by Edmund Husserl and Martin Heidegger. They consider the history of human cognition as a series of epochs of 'unconcealment' when the integrity of each one is provided by reconciliation, interaction and interpenetration of sundry practices.

Correspondingly, the epistemological accounts of the Ptolemy-Copernicus breakthrough miss a substantial point: the alternative ontologies of Ptolemy (mathematical astronomy) and Aristotle (qualitative physics) arose out of diverse social practices. Ptolemy's wayward ontology was grounded in pagan religious meditative practices (the divine mathematics of the Skies), and Aristotle's solid worldview was entrenched in sensual and object-manipulative ones (the mundane physics of the Earth). Copernican and Ptolemaic congenial programs were realizing the radically different from each other ways of astronomy and physics stout reconciliation.

For many centuries, the dualism had been 'frozen up' in the Aristotle-Ptolemy pagan cosmology. Following Edmund Husserl's studies, Koyrè maintained that modern science has succeeded in overcoming the split, inherent in Aristotelian science, between Earth and Space. The Christian canon Nicolaus Copernicus blew the cosmology up from within. The explosion was performed in a quite different (monotheistic) socio-cultural context, and, as a result, Galileo set up a throng of corresponding research practices aimed at the efficient reconciliation of mundane physics of the Earth with sublime mathematics of the Skies.

Correspondingly, the Copernican Revolution is envisaged in the context of intense interaction and interpenetration of Aristotelean and Ptolemaic sophisticated research practices. Eventually, the Aristotle – Ptolemy pagan cosmology could not help but be exposed to repeated cogent attacks during the Middle Ages since it apparently confronted the renowned principles of monotheism, not admitting the impervious demarcation line between the celestial and mundane realms. All the opposite worlds should have one and the same Creator. Commencing with the unification, Copernicus, in effect, paved the way for the descent of mathematics from Heaven to Earth and the spread of natural philosophy from Earth to Heaven.

Key words: Kuhn, Foucault, Rouse, Latour; Husserl, Heidegger, Sheffer, Schaffner, Kitcher, Longino, Copernican revolution, social facets, practice; Aristotle; Ptolemy; monotheistic Weltanschauung; Copernicus; Galileo; Newton.

I.1. Epistemological Accounts of the Copernican Revolution: Advantages and Disadvantages.

Why did bold and abstruse Copernican research program squeeze out refined and entrenched Ptolemaic one commencing the first scientific revolution? – The basic *epistemological* approaches to broaching the subject and solving the stubbornly recalcitrant problem are commonly laid out by the following prevalent options.

(I) Guileless inductivist version (William Whewell); **(II)** classical falsificationist version of Karl Popper; **(III)** intricate conventionalist version of Pierre Duhem;**(IV)** sophisticated falsificationist version of Imre Lakatos and Elie Zahar; **(V)** promising yet insufficiently advanced and controversial social-psychological version of Thomas S. Kuhn.

Nevertheless, the deft explanations for the ultimate reasons for Copernicus's triumph over Ptolemy seem insufficient in the light of the following plain counter-arguments.

(I) Guileless *inductivist* account famously turns out to be especially squalid since the theories from both competing research projects – that of Copernicus and Ptolemy – practically *equally* deviated from the available observational data (see, for instance, Kuhn 1977). It is a platitude that still the notorious Buridanists, in their controversy with Aristotle, contended that, on the unsteady ground of astronomical observations, "it is impossible to assert definitely whether the Earth or the stellar sphere moves" (Oresme; quoted from Kokowski 1996).

(II) Classical f*alsificationist* explanations of the substantial causes of Ptolemy's defeat are commonly reduced to the following two perspicuous options.

(II. a) According to the austere one, Ptolemy's deft theory was illicitly *irrefutable* and therefore unscientific while Copernicus's superlative theory was just the opposite. Ptolemy's ingenious heuristic was blatantly *ad hoc*. *Any* odd celestial fact could be aptly accounted for in retrospect by immense multiplying the intricate paraphernalia of heterogeneous epicycles, epicyclets, deferents, equants, and so forth.

Regrettably, the 'unrestricted proliferation' of manifold epicycles in Ptolemaic whimsical astronomy constitutes a wonted 'historical myth' (Gingerich 1997). In actual research practice, to compensate for the flagrant equant triumphal stave off, Copernicus had to insert a *new* species of no less stale epicycles. Eventually, Ptolemaic 'antediluvian' program transpired to embrace fewer epicycles than the 'bold revolutionary' Copernican one.

(II. b) According to the second, more sophisticated version (Popper [1935], 2002), both competing theories were practically equally lame for a sufficiently long time. But eventually, the smashing blow of the 'critical experiment' relentlessly refuted Ptolemy, buttressing Copernicus at the same time. Though when did this textbook miracle happen? Alas, historians of science have not come to a consensus yet. Maybe in 1616, when lucky devil Galileo had miraculously discovered the cycles of Venus?

Unfortunately, the generally-accepted claim that Galileo had successfully predicted the unexpected phases of Venus again is a sought-after historical lapse (Ariew 1989). One can rightfully sympathize with Galileo's scathing critics among the faithful Aristotelians decidedly refusing to take seriously the astonishing observational data produced by a newfangled telescope with rather dubious and murky work principles (Kuhn 1957; Feyerabend [1975], 2010; Chaunu 1984).

(**III**) Due to shrewd *conventionalist* approach, one cannot dare to make the final choice between the competing theories only grounded on sheer empirical considerations (Duhem [1906], 1954). Eventually, one theory surely fits better than the other since it is more 'simple,' 'beautiful,' 'coherent,' 'economic,' etc. Yet I. Lakatos and E. Zahar many a time and oft retorted that the alluring 'myth of simplicity,' for instance, was time and oft dispelled by the painstaking labor of the historians of science in variety of case-studies (Lakatos & Zahar 1974, p.362). Furthermore, such theory-choice situations are common for the history of science, when one of the competing theories is simpler than the other indeed. However, this 'other' is in better agreement with the available experimental data at hand or better reconciles with other respectable theories (see the examples in Kuhn 1977). Which theory should be *definitely* chosen? The same is true for other often misty conventionalist criteria that in common research practice function as values only and not as definite and effective criteria of choice.

(**IV**) The sublime SRP (scientific research programs) methodology (Lakatos 1970) enunciates that the pivotal problem of philosophy/methodology of science is to proffer a *normative* appraisal of scientific theories. A sturdy and objective *appraisal* of a scientific change is surely a *normative* problem and whereupon belongs to analytical philosophy. However, a stout *explanation* of the scientific change – of the elicited true reasons of acceptance and rejection of the theories involved – is predominantly a psychological/sociological conundrum.

Certainly, the Copernican distinctive program showed itself as a 'theoretically progressive' one. Not by chance this hallmark had been gladly anticipated by one of Copernicus' faithful disciples:

> "Aristotle says: 'That which causes derivative truths to be true is most true' (Metaphysics 993b 27-27). Accordingly, my teacher decided that he must assume such hypotheses as would contain causes capable of confirming the truth of the observations of previous centuries, and such as would themselves cause, we may hope, all *future* astronomical predictions of the phenomena to be found true" (Rheticus, Narratio prima,; quoted from Kokowski 1996, 39).

The Copernican fine program had theoretically accounted for many important 'novel facts' not observed before - star parallax, for instance. Though *actually*, the Copernican program started to make the reliable empirical progress only with Newtonian mechanics.

> "It was only with Galileo's, Kepler's, and Newton's modifications and supplementations that Copernicus's theory began to be seen as a clear successor to Ptolemy's." (Fuller 2020, p.4).

The 'Ptolemy-Copernicus' inextricable transition had been masterly reconstructed within the dazzling methodology of Imre Lakatos. In our view, one should not nitpick to petty lapses that are common for any pioneering treatise of this kind. Yet the crucial point is a *matter of principle*. One should never forget that the ultimate aim of the SRP methodology is to set an *objective and incontrovertible appraisal* of scientific change but not the humble *explanations* of its *actual* reasons (Lakatos & Zahar 1974, p.381).

Within the shadowy frame of the Lakatos-Zahar elevated approach, it is quite easy to admit that all the subtle content of the Copernican bold program could be contrived even by Aristarchus of Samos. But why it had *not* happen? And why did these sublime designs not have any significant impact on Ptolemy and his numerous followers? In our view, the abovementioned predicaments are connected with one and the same notorious reason – *lack of proper scrutiny of the subtle interconnections of cognitive and social aspects of scientific research.*

All those sublime (I) – (IV) epistemological models are defiantly one-sided ones. As a matter of fact they ignore to varying degrees the relentless hallmark that science constitutes a substantially '***community effect***' (Russ Payne). Regrettably, the epistemologists adamantly hold a 'great genius' vision of the progress of science. According to the romantic standpoint, the startling insights of very peculiar individuals, the 'chosen ones of the gods,' are what actually drives science forward. Though this is an apparently distorted picture. The great luminaries like Newton, Maxwell or Einstein can only *start* a revolution in scientific thinking at best – but only when a broader community of researchers relentlessly prepared the field. And sedulously created the conditions for the growth of the seeds of a revolution in thinking. Eventually, the history of science needs to be comprehended in new terms – in terms of *how* these broader communities progress to the point where revolutionary thinking is called for. The celebrated 'great insights and discoveries' never take place in a social vacuum. Revolutions in thinking can not happen until the scientific community is convinced that the old paradigm (program, tradition) is broken. Only when this does really happen, appropriate alternatives to the old and entrenched paradigm are advanced. The countless stories of unrecognized geniuses, beginning from Aristarchus of Samos, are the most convincing argument in favor of the above.

I.2. Thomas Kuhn's Intricate Yet Budding "*Structure of Scientific Revolutions.*"

(V) Copernicus was Kuhn's paradigmatic case of all paradigms (Heilbron 1998, p.508). Due to T.S. Kuhn's distinctive accounts (Kuhn 1957;1963, p.367), Ptolemy's sophisticated astronomy was going in 1543 through a depressive state of 'paradigm-crisis' that constitutes, according to the celebrated epistemological doctrine (Kuhn 1962), the indispensable 'prelude' to a scientific revolution.

> "Even Copernicus's more elaborate proposal was neither simpler nor more accurate than Ptolemy's system. Available observational tests [...] provided no basis for a choice [...] Ptolemaic astronomy had failed to solve its problems; the time had come to give a competitor a chance" (Kuhn [1962], 1970, pp.75-76).

Nevertheless, as Lakatos & Zahar (1974) shrewdly parried, how many scholars had perceived this horrific 'community crisis' that, withal, lasted for more than a *thousand* years? Herewith one of the celebrated historians of science persevered that in the Copernican paradigmatic case Kuhn oddly envisions 'a scandal where there was none' (Gingerich 2004).

And if Kuhn's meticulous analysis of the fine 'structure of scientific revolutions' is ultimately applicable to the Copernican case, why so *few* scholars had deliberately buttressed Copernicus before Kepler, Galileo, Descartes and Newton?

Further, according to Thomas Kuhn and his adherents, a scientific revolution – in a perspicuous analogy to political revolutions, such as the 1789 French revolution or the 1917 Russian ones – is taken as resolutely and completely displacing the '*Ancien Régime*' research traditions in science (Westman 1994). Via such relentless breakthroughs, new mature research traditions are so *radicalized* that they become irreversibly "*incommensurable*" with their mature predecessors. It leads to a bizarre standpoint that the great luminaries like Newton, Maxwell, Darwin, Bohr or Einstein, being the instigators of revolutionary breakthroughs, should be taken as convinced adversaries of the 'creative dialogue' between the 'old' and the 'new' paradigms.

However, such a bizarre tenet seems to us rather unilateral since it apparently exaggerates the revolutionary facet of a profound scientific change. Likewise, Copernicus rather aptly *modifies* than categorically *rejects* Aristotle's and Ptolemy's stout principles. Moreover, in many important respects he can be envisaged just as a discreet reformer of generally accepted astronomy rather than an ardent revolutionary intent on its overthrow. For instance, Copernicus discreetly modifies famous Aristotelian principle that an object can take part only in a single natural motion. He aptly argues that a more complex principle is required to accommodate falling bodies on a rotating Earth (Goddu 2010). Kuhn himself had to gingerly notice that Copernicus worked in the wake of Ptolemy, from whom he derived his methodology, his conceptual tools and even the structure of his major work (Omodeo 2017, p.93).

Likewise, in his 1949 "Autobiography" Einstein famously noted that "in the theory of relativity we are undoubtedly not seeing a revolutionary act, but a natural development of those trends that have developed over the centuries."

Not by chance in his blunt "*The Revolution That Didn't Happen*?" Steven Weinberg called Kuhn's description of scientific revolutions 'seriously misleading,' insisting that changes in understanding nature "have been *evolutionary, not revolutionary*" (quoted from Holton, 2017, p.32).

However, Kuhn certainly developed some really profound sociological insights which were integral to his abovementioned analysis of scientific knowledge (Bloor 2017). Regrettably, his 'sociality' is too narrow since it is reduced to social-psychological reasons of a paradigm crisis. What is left overboard is the full, integral **social** analyses of the origins and the necessary outcome of a scientific revolution.

Though the 'world-in-itself' does not change, the 'phenomenological world' of a scientist might change during a scientific revolution (Hoyningen-Huene 1993, 32ff). The world that undoubtedly changes is the '*social world* of the scientist' (Bloor) that contains such things as the subtle standards of reasonable agreement between theory and experiment. Hence in the scientific revolutions we are not, primarily, dealing with phenomenological changes but with *procedural* and *behavioral* ones. The changes span the standards that are supported by the profession as social institution defined as a set of social roles and statuses linked by conventions. So, they can be said to exist because 'they are believed to exist.'

Turning to the Copernican revolution, note that Kuhn soberly described the transition from Aristotle's mechanics to Galileo's one by stating that, before the transition, there were "no pendulums - only swinging stones" (Kuhn [1962], 1970, p.121). Though the change will be invisible if the attention is focused only on the swinging stone. To investigate the change it is necessary to consider *all* the persons and activity around the swinging stone. "It is necessary to attend to what these persons, collectively, do to, and with, the material object" (Bloor 2017, p.243). Thus, there is no other external criterion than the authoritative ***practice*** of scientific community itself.

Eventually, the subtle norms governing the correct usage of the word 'pendulum' ceased to be confined to ones that could be followed on the basis of merely visual observation. They embrace the accepted ways in which the object can become the subject to mathematical analysis. The correct application of 'pendulum' in particular theoretical scheme (see especially the second chapter) was now related to such components of the fundamental theoretical scheme of classical mechanics as 'force' and 'inertial system of reference.' – And, of course, such engineering concepts as 'escapement.'

Consequently, social reality is something created through a process of self-reference, and Kuhn's pivotal analytical concept, the 'paradigm,' has this character. "An achievement becomes a paradigm in virtue of being *regarded* as a paradigm, *treated* as a paradigm and *known* as a paradigm" (Bloor 2017, p.250).

But then the next question inevitably arises: what are the origins of the novel paradigm? – Kuhn's "*Structure of Scientific Revolutions*" seems to lack the definite answer reducing the contrivance of a novel paradigm to quasi-religious 'gestalt-switch.'

Note that intellectual space for the creation of sociology of scientific knowledge (SSK) was made by Thomas Kuhn himself. His '*Structure of Scientific Revolutions*' ([1962], 1970), in turn, was the application of Ludwig Wittgenstein's idea of '*form-of-life*' to practice of science (Collins 2017).

On the other hand, one cannot ignore a certain 'Hegelian slant' (Stachel) in '*Structure of Scientific Revolutions.*' Indeed, in Hegelian spirit, science cannot be reduced even to the best intentions of individual scientists, but is an '*entity in itself.*' Hence a mathematical theorem is not true because someone got an idea and then successfully applied the universal laws of logic or math, thereby proving the theorem. Rather, according to one of French Hegelians, the 'truth' of the theorem is in its very demonstration, which represents a necessary movement within the structure of science itself.

> "The true meaning of a theory is not in what is understood by the scientist, as he [Jean Cavailles] wrote in "*On Logic and the Theory of Science*," but in conceptual becoming that cannot be halted' (Stachel 2017, p.318).

Respectively, the task of a historian of science is to elicit the constitution of truth as a historical concept within an era, rather than to investigate what was believed to be true in that era. In that respect,

> "Tom [Kuhn] understood history of science as history of ideas, in the strong sense that ideas were the active agents in history" (Norton Wise 2017, p.347).
>
> Forsooth, their [paradigms] 'very special efficiency' depends on
>
> "the unparalleled insulation of mature scientific communities from the demands of the laity and of everyday life."
>
> Even in conditions of crisis for a paradigm,
>
> "technical breakdown still remains the core of the crisis" so that "external factors" were secondary (Kuhn [1962], 1970, p.164, p.69).

Kuhn's paradigm-shift theory represented in many respects a generalization (an 'epistemologization' – Omodeo 2017, p. 74) of Koyrè's conception of scientific revolutions. For instance, in an entry on the History of Science, for the International Encyclopedia of the Social Sciences (1968), Kuhn pitted the "Duhemian school," in which he included Alexandre Koyrè, against the Marxist historiography.

For instance, Galileo's mathematical physics famously was a kind of Platonic contemplation of the numbers and geometries hidden behind the natural phenomena. Just as Galileo was a committed Platonist (Koyrè 1943, p. 424), the first scientific revolution was a "spiritual revolution" (Koyrè 1943, p.403). And as Yehuda Elkana correctly put it,

> "[Koyre's studies became a paradigm for history of science as history of disembodied ideas" (quoted from Omodeo 2017, p.74).

Furthermore, in the epoch of the first scientific revolution, mechanics advanced independently of astronomy, regarding its Archimedean legacy as well as material, technical and social roots. The advancement of statics from ancient times to the Renaissance

> "illustrates in an excellent manner the process of formation of science generally [...] These beginnings point unmistakably to their origin in the experiences of the manual arts" (Mach 1942, p.89).

In that sense, celestial physics was an artful extension of terrestrial mechanics to astronomy. Nevertheless, according to Koyrè, eventually the first scientific revolution originated from a cosmological turn (see the next parts of the chapter).

Moreover, Thomas Kuhn's thorough "*Mathematical vs. Experimental Traditions in the Development of Physical Science*" (1976) revised the famous 'Merton thesis' by restricting its significance only to those sciences that were not proper 'science.' He stressed that whereas technology and economy had a significant impact on the emerging experimental sciences, they did not affect astronomy, mechanics and other advanced 'classical sciences.'

However, one should not ignore Rudolph Carnap's letter to Thomas Kuhn written just after the first 1962 publication of "*Structure*"

> "I found very illuminating the parallel you draw with Darwinian evolution: just as Darwin gave up the earlier idea that the evolution was directed towards a predetermined goal, men as perfect organism, and saw it as a process of improvement by natural selection, you emphasize that the development of theories is not directed towards the perfect true theory, but is a process of improvement of an instrument" (quoted from Engler & Renn 2017, p.146).

To sum up, Kuhn's "*Structure of Scientific Revolutions*" is a fortuitous attempt to find a compromise, to ***reconcile*** two opposite trends of history

&philosophy of science typified by Wittgenstein (forms of life) and Koyrè (internal history of science). Now one can comprehend Koyrè's remark in his post-Structure letter to Kuhn:

> "You have brought the internal and external histories of science, which in the past have been very far apart, together" (quoted from Stachel 2017, p.322).

Yet one should be clearly aware that that Kuhn's effort was merely a 'first approximation.' Using Yehuda Elkana's term, Kuhn's image of science does not take into account many dimensions of the advancement of science that since 1962 have become central to modern history & philosophy of science debates.

In that sense, Kuhn's compromise between cognitive and social facets consisted first and foremost in that he proffered not only epistemological but also a *sociological ideal model* of the progress of science as well, artfully ignoring those social aspects that he considered unimportant 'in the first approximation.'

> "If history is to rise above a mere collection of anecdotes, it must be written from some *point of view* and with some *unifying theme*. It is here that the philosopher has a contribution to make by supplying some *distinctive perspective*, such as Kuhn's view about paradigms, normal science and revolutions" (Shea 2017, p. 172, my italics).

Those aspect should be consequently taken into account in future more refined and exact models in direct accordance with sundry practice of theoretical physics. In that respect the fact that he had obtained his first PhD in theoretical physics turned out crucial.

Using Jed Buchwald's apt words,

> "Kuhn remains relevant today precisely because he insisted on probing *beneath* the surface of scientific discourse to *reveal* the *unstated* but powerfully operative *practices* and beliefs that characterize a group" (Buchwald 2017, p.161).

Furthermore, in his influential book "*Making 20th Century Science:*

How Theories Became Knowledge" (2015) Stephen Brush, on the basis of reception studies of STR, GTR, quantum theory and Darwinism, discloses that mature theories are accepted due to historically random combinations of successful predictions and better explanations of 'old' facts. Why have scientists accepted mature theories? – Brush discovers that historians and epistemologists neglect this question. Historians focus on the genesis of mature theories, or on the peculiar circumstances of their acceptance, without interrogating scientists' rationales for accepting them. Epistemologists are commonly too preoccupied with the highly controversial question of why scientists *should* accept mature theories to examine *how* they actually do.

On the contrary, Brush unfolds four ideal types of potential explanations of how a novel mature theory successfully wins scientists' allegiance. (1) It might successfully anticipate hitherto unknown scientific facts (novel prediction); (2) it might provide a compelling account of previously known theories and facts (explanatory power); (3) it might be more slim and 'beautiful' than its competitors (aesthetics);(4) it might be favored by social, cultural, or professional factors (social construction).

Note that Brush comes to a distinctive conclusion that scientists often accept mature theories first and foremost for their explanatory power. Ipso facto he attacks the 'hypothetico-deductive method' and highly praises social constructionists for scrutinizing scientific practice.

To sum up, Thomas Kuhn was one of the first to realize that *the further advancement of philosophy of scientific revolutions should consist in reconciliation or harmonization of the so-called cognitive and social approaches, 'internalism' and 'externalism.'* However, his own social view of science was too narrow being in fact reduced to a kind of social-psychological account. Hence it should be stressed that, *since Kuhn, the fruitful advancement of history &philosophy of science could proceed only by preserving the compromise.* All these not forbids the advancement of any of the components of the compromise but only on the condition of the restoration of their harmonic whole.

Thus, in my view the evolution of Kuhn's paradigm concept started from the social constructionist standpoint.

I.3. Bruno Latour's Sociology of Scientific Knowledge: in the Wake of the '*Structure.*'

In the second half of the XXth century, probably after the end of the Second World War (with its concentration camps, massacres and Hiroshima and Nagasaki) there occurred an urgently felt need to comprehend science as human enterprise not dissociated from its social, economic and cultural conditions. And

> "only the decreasing credibility of the normative, rationalist accounts of science in the tradition of analytical philosophy has opened up the possibility to conceive the role of social structures for the scientific enterprise not only as belonging to the context of application but also to the context of the constitution of the validity of scientific knowledge. Also the presently much debated role of experimental cultures in the history of science shows traces both of a concern with the present state of science, which is often characterized by large-scale experimental systems which evolve semi-independently from theoretical endeavors, and of the lost faith in the centrality of language and logic for understanding science as it was proclaimed by analytical philosophy" (Renn 1996, p.2).

Latour and Woolgar's 1979 persistent study decidedly broke away from unilateral positivist view of advancement of science as a sheer asocial process capable of revealing universally valid truths regarding the natural world. On the contrary, they stubbornly took scientific knowledge as an *artificial* product of different social, political, and economic *interactions.*

However, at first the pivotal tenet of modern SSK consisted in the assertion that it is scant to scrutinize the language (of science) through an analysis of words and meanings isolated from the *pragmatic* situations in which they are used (Lynch & Woolgar 1990). This applies not only to words but to methodological rules and recipes and theoretical propositions as well. Although such formulations find their particular applications in real scientific accounts, they are relentlessly fetishized in epistemological writings on basic scientific 'ideas' and apt experimental 'methodology.'

> "For him [for Popper], 'science' was the name of a potentially universal attitude rather than a set of socially specific techniques [...] Whereas Popper basically saw the scientific attitude and liberalism as alternative expression of the 'open society,' Kuhn traced science's unique epistemic strength to its authoritarian if not totalitarian structure" (Fuller 2003, p.2).

Forsooth, sociologists have no need to meticulously clarify everyday language, insofar as 'everyday' or 'ordinary' actions are the primary subjects of empirical study. Instead of asking, 'what do we really mean, in various contexts, by 'representation'?," the studies commence by asking "What do the *participants* take as representation?"

Following the trend, philosophers, historians, and sociologists of science are now strenuously questioning the privilege traditionally assigned to the verbal statement or 'proposition,' devoting instead attention to multifarious forms of representation.

The corresponding studies unfold an abundance of ways to disclose that the particular 'representations' have little determinate meaning or logical force aside from the complex activities in which they are situated. Eventually the startling topic of 'representational practices in science' was related to the general issues in the sociology and philosophy of science (Lynch & Woolgar 1990). Herewith a strong emphasis on the heterogeneity and discontinuity of representational formats, and on the local and contextual ground of their production, has become a common refrain in 'postmodernist' critiques of traditional accounts on science (Derrida 1970; Foucault 1969).

However, the conspicuous trend under consideration was epitomized in Barnes & Bloor's renowned 'Strong Program' in the sociology of knowledge. A pivotal aim of the program is to contrive sociological explanations of *all* scientific 'beliefs,' regardless of the truth or falsity assigned to them. The 'strong program' treats the 'contents' of scientific knowledge as reified products of historical and contemporary stark controversies, sedulous partisan genealogies of scientific facts, and outrageous negotiations over credit for discoveries.

On the one hand, socio-historical arguments on how scientists' theoretical commitments reflected social 'interests' were 'interesting and suggestive' (Lynch & Woolgar).Though on the other hand they begged many inextricable questions on the nature of such 'interests' and how they work themselves into research settings (Woolgar 1981).Furthermore, according to Paul Tibbets, Bloor's and Barnes's constructive arguments are reflexively undermined by incongruent causal and realist assumptions (see also Laudan 1981; Woolgar 1983).

Respectively, bold

> "hypotheses about changes in the mind or human consciousness, in the structure of the brain, in social relations, in 'mentalitè's,' or in the economic infrastructure which are posited to explain the emergence of science or its present achievements are simply *too grandiose*, not to say *hagiographic*, in most cases and plainly racist in more than a few others...The idea that a more rational mind or a more constraining scientific method emerged from darkness and chaos is too complicated a hypothesis. It seems to me that the first step toward a convincing explanation is to adopt this a priori position" (Latour 1990, p.19).

The key point is that the Grand Dichotomy with its 'racist hubris' certainty should be replaced by many uncertain and often unexpected divides (see also Goody 1977). According to Bruno Latour's apt analogy, the notorious divide between prescientific and scientific culture is merely a common border – like between Tijuana and San Diego. It is installed by the border guards dominated by the bureaucrats; yet it does not represent any *natural* boundary. Eventually, all these sublime Grand Dichotomies cannot provide any reliable explanation. On the contrary, they are themselves the things to be properly explained (Latour 1983). But in providing reliable and stout explanations one has to steer a course that can lead out of a perfunctory relativism by eliciting a number of simple, empirically verifiable causes. One needs to arrive at more *mundane* explanations than that of a great divide in human consciousness.

For instance, Kuhn famously describes the shift from the 'old' paradigm to a 'new' one as a 'conversion process.'

> "The conversion experience that I have likened to a gestalt switch remains [...] *at the heart* of the revolutionary process. Good

> reasons for choice provide motives for conversion and a climate in which it is more likely to occur. Translation may, in addition, provide points of entry for the neural reprogramming that, however inscrutable at this time, must underlie conversion. But neither good reason nor translation constitute conversion, and it is that process we must explicate in order to understand an essential sort of scientific change" (Kuhn [1962], 1970, p.204).

However, in Joseph Agassi's distinctive terms, it can be asserted that

> "First and foremost, the very idea that scientific research rests on conversion, that conversion of import for research may be vague and quite general, yet later become more specific. We have so many examples of researchers who tried their hands in diverse directions, who developed ideas which were quite contrary to their deepest convictions, who developed ideas of their opponents in the spirit of self-criticism, in the spirit of criticism, or both. It is time to stop writing the history of scientific research while using religious metaphors" (Agassi 1983, p.379).

As for Barnes & Bloor, worst of all in order to explain science one has to kneel before one peculiar science, that of economics. Thus, ironically, many 'materialist' studies of the profound origins of science are in no way material. They defiantly ignore the precise practices and subtle craftmanship of knowing. Hence

> "it seems to me that the only way to escape the simplistic relativist position is to avoid both 'materialist' and 'mentalist' explanations at all costs and to look instead for more parsimonious accounts, which are empirical through and through, and yet able to explain the vast effects of science and technology" (Latour 1990, p.21).

In sum, Bruno Latour finds all the explanations in terms of social groups, interests or economic trends convincing only when they proffer a specific *mechanism* to sum up 'groups,' 'interests,' 'money' and 'trends.'

Turning to the conspicuous case of the first scientific revolution, one can conclude that the rationalization that took place was not only and not so much of the mind, of the eye, of philosophy but of the sight. For instance, why was perspective such an important invention? "Because of its logical recognition of internal invariances through all the transformations produced in spatial location" (Ivins 1973, p. 9). Hence in the West, even if the subject of the printed text were guilelessly unscientific,

> "the printed picture always presented a rational image based on the *universal laws of geometry*. Hence the Scientific Revolution all appearance owes more to Albrecht Durer than to Leonardo da Vinci" (Edgerton 1980, p.190).

In this sense

> "Antonello's St. Jerome is an apt paradigm of a new consciousness of the physical world attained by Western European intellectuals by the late XVth century. This consciousness was exhibited by such artists as Leonardo da Vinci, Albrecht Durer, Hans Holbein and others who "developed a sophisticated grammar and syntax for *quantifying* natural phenomena in pictures"" (Edgerton 1980, 189).

That is why Latour resolutely rejected the so-called 'materialist' explanations that apply 'infrastructures,' or 'markets' or even 'consumer needs' to account for the advancement of science. On the contrary, it is the visual construction of something like a 'market' or an 'economy' that cries for explanation. Moreover, one commonly takes for granted that there exist, somewhere in society, powerful macro-actors that naturally dominate the social scene: Corporation, State, Material Productive Forces, Cultures, Imperialism, etc.

But the problem is that these conspicuous entities could not exist at all. A 'state,' a 'corporation,' a 'culture,' an 'economy' et al. are the result of a punctualization process that obtains a few indicators out of many traces. In order to exist these entities have to be summed up once upon a time and somewhere. Far from being the key to comprehension of science, the abovementioned entities are the very things the new comprehension of science should explain in its turn. The large – scale actors to which the partisans of the Strong Program are keen to attach 'interests' are immaterial in practice as long as precise and empirically verified mechanisms to explain their origin or extraction and their changes of scale have not been proposed.

> "To take the existence of macro-actors for granted without studying the material that makes them 'macro,' is to make both science and society *mysterious*." (Latour 1990, 56).

Furthermore, many dazzling accounts were proffered to hackup the history of science with the history of capitalism. For instance, many efforts have been made to display the scientist as capitalist. According to 1990 Bruno Latour, these startling efforts (even including Latour & Woolgar 1979, chapter V) were 'doomed from the start,' since they took for granted the notorious division between mental and material factors. Correspondingly, there is not a history of engineers, then a history of capitalists, then one of physicists, chemists, biologists, and mathematicians, then one of economists and accountants. On the contrary, there is a single history of these important centers of calculation.

> "It is because all these inscriptions can be *superimposed, reshuffled, recombined,* and summarized, and that totally new phenomena emerge, hidden from the other people from whom all these inscriptions have been exacted" (Latour 1990, p.60).

Finally, two tempting attempts have been made to relate the structure of cognitive abilities to social structure. The first one applies Kuhn's budding paradigms (Barnes 1982), while the second adheres to Wittgenstein's thought-provoking 'language games' (Bloor 1983).

The conspicuous attempts are interesting but they still try to answer a question which Latour strongly recommends to reject: how cognitive subtle abilities are related to our societies. The key point is that the inextricable question (and the various answers) accept the idea that the stuff our society is made off is somehow *different* from that of our sciences, our images, and our information. Yet

"we are dealing with a single ethnographic puzzle: some societies – very few indeed- are made by capitalizing on a larger scale. The obsession with rapid displacement and stable invariance, for powerful and safe linkages, is not a part of our culture, or 'influenced' by social interests: *it is our culture.* Too often sociologists look for *indirect* relations between 'interests' and 'technical' details. The reason of their blindness is simple: they limit the meaning of 'social' to society without realizing that the mobilizing of allies and, in general, the transformation of weak into strong associations, is what 'social' also means. Why look for far-fetched relations when technical details of science talk directly of invariance, association, displacement, immutability, and so on?" (Latour 1990, 64)

To sum up, *the sociological attempts to 'correct' the one-sidedness of Kuhn's reconciliation project, though correct as a general trend, turned out to be too vague and abstract.* However, a promising direction had been crystallized embodied by Michelle Foucault's reflection on 'practice.'

I.4. Foucauldian Concept of Practice: Overcoming the One-Sidedness of '*Structure.*'

As is well known, just at the end of the 70s, the transition from the classical Mertonian paradigm to the so-called 'strong program of the sociology of knowledge' had happened. The transition first and foremost consisted in that not merely patterns of behavior and values, but *scientific knowledge* itself became the subject of relentless sociological analysis (see, for instance, Mulkay 2015). Eventually, the Mertonian account is comprehended now not as an empirically adequate picture of science, but rather as a useful ideology concocted to promote the institutional independence of sciences.

Even though many scholars maintain that the blunt transition was caused by purely empirical reasons, associated with the increasing proportion of inextricable empirical material to survey, it seems to me that this is not entirely true. Of certain importance transpired also the changes in epistemological vistas caused by the assimilation of the advances of world philosophical and sociological pundits.

This paragraph aims to exhibit the impact of Michel Foucault's distinctive writings on the sociology of scientific knowledge. In particular, the purpose of the work is not only to acknowledge that Foucault's work influenced the formation of the modern sociology of knowledge but also to underscore the alluring facets of this influence, which make it possible to salve the number of stale problems facing the domain. To do it, and to elicit the pivotal ideas that are of a non-random nature and permeate all his work, we will contemplate the evolution of Foucault's ideas first, highlighting its main milestones, and examining the results that he received gradually during the transition from one stage to another.

Our research is inspired by the following Foucauldian dictum:

> "In this part of the study of the penitentiary system, as in my early works, the target of analysis wasn't 'institutions,' 'theories' or 'ideology,' but *practices* [...] being understood here as places where what is said and what is done, rules imposed and reasons given, the planned and the taken for granted meet and *interconnect*" (quoted in Burchell 1991, p.75; my italics).

Accordingly, the following basic research problems run through Foucault's manifold writings (Power 2011, pp.35-56):

- the problem of the inextricable nature of discourse and the subtle archeology of knowledge;
- the historical method adequacy problem in the sociology of knowledge;
- the problem of inextricable social action and its multifarious actors;
- the problem of the abstruse relationship between knowledge and power.

Before proceeding to an inquiry of how these inextricable problems gradually arose and intertwined in his work, let us make the following reservation. Michel Foucault himself did not consciously strive to build an integral and coherent *system* of the sociology of knowledge; therefore, the lion's share of information about his creative ideas is proffered not only by his stupendous books but by profound and thought-provoking interviews that he gave to multilateral, mostly French, magazines as well (Foucault 1980).

It is commonly held that the first and especially widely known work of Foucault, which made him a name, was his masterpiece "*The History of Madness*" (Foucault 2009), first published in 1961 as "*Folie et Deraison: Histoire de la folie a l'age Classique*."

Already in this work one of his pivotal 'guiding ideas' can be traced: the stiff tendency to withstand the enlightenment and functionalist concepts of scientific progress. In this book, he convincingly exhibited that the so-called 'progress' in psychiatry consisted in the relentless and callous exclusion from the sphere of the scientific discourse of experiences, opinions, and values of vast groups of people – the so-called 'insane,' whom the pundits simply ceased to regard as people at all. Foucault convincingly demonstrates that in pre-modernist societies, on the contrary, the demarcation line between reason and madness was not as sharp and impenetrable as in modern societies; they did not strive to carry it out with the same ardor as in modern times. Incidentally, Foucault famously distinguished three basic stages in the development of any scientific discipline - pre-classical, classical, and modern (for more details see Power 2011, p.36).

In the blissful pre-classical era, which lasted until the middle of the XVIIth century, the notion 'madness' was not a social category at all. But everything changed after the opening of a general hospital in Paris in 1657, in which special wards for the insane people were set up. The end of the eighteenth century until the modern era was in 1794 marked by the establishment of a psychiatric hospital and a rigidly delineated dividing line between those patients who can and should be properly treated, and the 'hopeless' ones who should be decisively isolated from society.

It is at this point that a new era in the relationship between madness and medical science starts. In his alluring chef-d-oeuvre, "*The Birth of the Clinic*" (Foucault 1994), first published in 1963 as "*Naissance de la Clinique: une archeologie du regard medical,*" the author is even more obsessed by revealing the hidden rules that make possible the emergence and 'constituting' of authorized knowledge systems. Even the human body itself becomes the locus of diagnoses and expert ratings. Note that Foucault takes a very restrained position about common, standard humanism with its inextricable struggle with the enslavement of the patient by a group of expert doctors and soberly does not try to oppose one stage of the advancement of medical knowledge to another.

On the contrary, the peculiarity of his approach consists merely in an attempt to display *objectively*, without excessive sentimentality, how the practices of considering the human body in the unity of medical and legal prerequisites are gradually changing. This specificity becomes even more apparent in his next epoch-making work - in the book "*The Order of Things: An Archeology of the Human Sciences*" (Foucault 1994), first published in 1966 as "*Le mots et les choses: Une archeologie des sciences humaines*." It is this work that presents in the fullest form the basic ideas of his abstruse historical sociology of knowledge.

(It is no coincidence that in the same year the pre-eminent book by Peter L. Berger and Thomas Luckmann "*The Social Construction of Reality: A Treatise in the Sociology of Knowledge*" was published. The book set up a dialectical view on society both as 'objective reality' and as 'subjective' one).

The gist of Foucault's work is to reveal the strangeness and randomness of the emergence of knowledge practices in the historical arena, to broach the subject of how these practices function, and to examine the type of order that they constitute. As before, in drastic opposition to common historians of ideas, Foucault is focused on revealing the *hidden conditions* of the possibility of knowledge and the randomly emerging relationships between their parts that underlie various scientific classifications. Seen from this perspective, his earlier abovementioned work on insanity and medicine can be regarded as historical case studies scrutinizing how different '*regimes of truth*' emerge, each in its field. It is no coincidence that worldly-wise scholars could not fail to note the significant affinity between the writings of Michel Foucault and Thomas Kuhn (see, for instance, Hacking 1979). Nevertheless, unlike Kuhn's standpoint, Foucault's approach is much closer to Bloor's and Barnes's strong program for the sociology of knowledge, especially to its pivotal 'symmetry postulate' in accounting for true and false beliefs (Barnes 1977; Bloor 1991).

Foucault, on the contrary, is interested not in truth as such, but in the social and especially institutional *historical conditions* under which authorized, statements are constituted as 'true.' Therefore, in this stupendous work, Foucault recounts three realms that underwent a particularly significant transformation at the end of the eighteenth century: biology, economics, and linguistics. Until that time, these domains of knowledge were purely empirical, being overloaded with all sorts of notorious primitive and superficial classification schemes that were emphatically objectivist. No attempts have been made to take into account the activity of the cognitive subject in the formation of representational practices.

Therefore, according to Foucault, the three disciplines epitomized a radical break with the past and the set up of a new '*episteme*' in the following two important respects.

(1) First, each domain of knowledge acquires an additional epistemological dimension in the form of a demand to penetrate the surface of things. For instance, there exist development processes behind the available living organisms. Behind the surface properties of money functions a whole system of production and reproduction of human well-being; and, finally, under the surface of grammatical rules are subtle mechanisms for transforming and adapting languages.

(2) Second, Foucault stresses that in each of the domains considered, the abovementioned 'gap' consists in focusing on a person as an object of research; in this sense, forsooth 'man is a recent invention' (Foucault).

In "*The Order of Things*" Foucault punctuates that the actors of modern social science are contrived and invented through knowledge-based practices that individualize their objects in ways that we today take for granted. Especially the author of "*The Order of Things*" contends that the modern host of academic disciplines, including sociology, is the result of the joint emergence of the empirical field, new (Kantian) forms of philosophical reflection of the subject of knowledge, and the growing weight of mathematics in the natural sciences.

In his next masterpiece, '*The Archaeology of Knowledge*' (Foucault 2002), first published as '*L'archeologie du savour*' in 1996, Foucault subjects his epistemic methods and the entire quirk conceptual apparatus to relentless philosophical reflection. In this case, he calls it 'archaeological.' Statements and propositions are the main units of analysis now, and their integrity, totality establishes discursive formations, which are the conditions for the possibility of thought and action. These specific discursive formations (or '*epistemes*') create the positions occupied by subjects in which the latter can authorize to speak. In this respect, the connection with the works of Marx (Lukács) and Freud (Lacan) is blatant (see Zizek 2009).

The staggering book '*Discipline and Punish*: *The Birth of the Prison*' (Foucault 1977), first published in 1975 as '*Surveiller et punir: Naissance de la prison*,' commences with genuinely Foucauldian dramatic juxtaposition of the two episodes. They date back to different historical epochs and particularly clearly reveal the depth and the historical gap that took place: scenes of relentless public execution and an outlook of prison as a social institution. The author applies this juxtaposition to posit how the barbarian infliction of pain has been gradually replaced by a subtle discipline of body submission - supervision, correction, and training in an enclosed space. Drawing on Bentham's 'panopticon' ideas, Foucault puts forward 'the constant possibility of supervision' as an overwhelming principle of power over the body and coordination of human behavior about other people.

Further, Foucault generalizes the principle of the panopticon to the entire system of organization of public life in modern society, which embraces diverse and extremely sophisticated surveillance practices. It is no coincidence that this work laid the foundation for the celebrated analysis of power in society; though power relations are 'diffused' in society, and not simply and primitively emanate from some power center, as it might seem at first glance. Note that the concept of 'knowledge-power' (pouvoir-savoir) reflects the fact that knowledge is stiffly embedded in the blunt practice of control, as well as in the corresponding forms of resistance to them.

It is equally important that the starting point of Foucauldian reflections on knowledge-power are not legal institutions, not administrative ones, but what Foucault refers to as the low archival foundations of local and specific types of disciplinary knowledge, which are consistently objectified in documents. Accordingly, Foucault is not craving to put up a 'theory of power' in the sublime spirit of modern refined political theory. On the contrary, he directs his research gaze to completely mundane, common, banal rules embedded in numerous, often everyday practices -

rules that determine what can be expressed, cognized, and done, by someone and for someone.

While Foucault's analysis of prisons, hospitals, and psychiatric wards seems intuitively fair, the situation drastically changes when he addresses the notorious issue of sexuality. Though the four-volume '*History of Sexuality*' (Foucault 1976-2021, the first volume '*An Introduction*' published in 1976), is not at all about 'sex' in the modern sense of the word, about the discourse of sexuality and, most importantly, about the inextricable hidden mechanisms for the production of socially acceptable statements about sexual behavior. Incidentally, in this work, Foucault speaks out against the popular 'repressive' hypothesis that power suppresses sex, developing instead rather profound and subtle considerations about the fine dialectic of the mutual influence and interpenetration of power and sexual relations.

As a result, Foucault's self-esteem of his work as a whole seems to be sufficiently fair.

> "My books are not treatises in philosophy or studies of history: at most, they are philosophical fragments put to work in a historical field of problems" (quoted from Burchill 1991, p. 74).

Foucault's most conspicuous achievement seems to be the conceiving of professional knowledge not as a given, but as an integral *part of certain practices* designed to determine and moderate human behavior. In particular, his contemplation of formal legal systems of knowledge as an ideal facet of the practice of controlling human behavior deserves particularly high applause. In this sense, Foucault's interest in micro-technologies of power diminishes the role of the modern state and its stiff structures.

According to Reiner Keller's sober dictum,

> "Foucault's fundamental achievement was to look at discourses as socio-historically situated 'practices,' and not as the development of ideas or lines of argumentation, and to 'liberate' discourse analysis from the specific linguistic issues...He addresses discourses as *battlefields*, as power struggles around the legitimate definitions of phenomena." (Keller 2011, p.46).

Returning to the problem of Foucault's influence on the so-called 'historical turn' in the philosophy & sociology of science, it should be noted that his ideas had a particularly significant impact not only on Kuhn himself but on his modern followers as well.

As we have already mentioned, according to Thomas Kuhn and his partisans, a scientific revolution – in a far-going analogy to tragical political events, such as the French revolution or the Russian ones – is taken as resolutely and completely displacing the 'Ancien Regime' scientific traditions (Westman 1994). And through the course of such relentless breakthroughs, novel mature theories are so *radicalized* that they are contrived to be '*incommensurable*' with their fundamental predecessors. This with necessity leads to a standpoint that the great scholars, being the instigators of revolutionary breakthroughs, may be pictured as convinced *adversaries* of the 'creative dialogue' between the research traditions including the thrilling case of the 'old' and the 'new' ones.

Such a strong tenet (and Kuhn's startling 'imagery of warfare' used to depict the reception of the new 'paradigm') punctuates the revolutionary facet of truly deep scientific change.

In my humble view, fierce controversies around the notorious 'incommensurability thesis' are tightly connected with another hallmark of Kuhnian epistemological doctrine. Indeed, many times Thomas Kuhn had mentioned that external factors might help decide the outcomes of some scientific revolutions (see, for instance, Sun worship in constructing Keplerian sophisticated planetary models, etc. - Kuhn [1962], 1970, pp.152-153).

Though he vigorously stressed the fact that during the most profound scientific revolutions – for instance, the Copernican one, - the scientists were responding first and foremost to the refined problems raised *within* science. What appealed to them in Copernicus's distinctive theory was its ability

> "to do away with ad hoc devices in Ptolemy's system (such as the equant), to explain key phenomena in a pleading fashion (the observed retrograde motion of the planets), and to explain away otherwise inexplicable coincidences in Ptolemy's system (such as

the alignment of the Sun and the centers of the epicycles of the inferior planets." (Bird 2018).

Positing himself over and over again as a 'pretty *straight internalist*,' Kuhn asserts that in a scientific revolution a worthy replacement (a 'new' theory) must retain *all* its predecessor's power to solve *quantitative* problems (Kuhn [1962], 1970, p.169). Whereupon one can conclude that scientific revolutions should necessarily bring with them an overall increase in *puzzle-solving power*; herewith the number and significance of the puzzles and anomalies solved by the 'new' paradigm should exceed significantly the number and significance of the 'old' one. For instance, the acme of the protracted Copernican revolution is epitomized in the creation of the differential calculus of Newtonian mechanics capable of solving the sophisticated problems of motion.

I.5. Joseph Rouse's Practical Hermeneutics: Pro et Contra

Furthermore, a fruitful attempt to apprehend the gist of Kuhn's concept and, successfully to overcome a number of its shortcomings, to advance this concept further based on Foucault's provoking writings is Joseph Rouse's research project. (See especially his startling paper "*Kuhn: The Philosophy of Scientific Practice*" (Rouse 2002, pp. 100-121).Rouse starts philosophizing by noting that Wittgenstein and Heidegger challenged conceptions of normative ordering as governance by explicit rules or norms. 'Rule-following' draws on 'agreement in form of life.'

According to Joseph Rouse, a crucial part of Kuhn's point, due to the influence of Wittgenstein and Heidegger, is to distinguish paradigms from *explicit* principles or rules. Viz., grasping a paradigm is certainly *not* like holding a belief or following a rule, but is instead acquiring a complex but flexible set of skills. These are, first and foremost, skills of recognition. Though this recognition presupposes more basic skills of doing math, applying concepts, inventing, building and manipulating apparatus, etc. Even when talking about theoretical understanding, Thomas Kuhn underscores the skillful use of concepts, models and symbolic generalizations as interconnected tools, rather than blunt beliefs or judgements that may result from their application. "*Scientific understanding is dynamic and practical*"!

Hence the main advantage of Kuhn's concept, which favorably distinguishes it from logical empiricism, is an attempt to describe science not as a stale system of principles and rigid, logically verified rules that bind it, but as a vivid *research activity* as such, a sheer '***scientific practice***.'

And the celebrated 'paradigms' should be comprehended not as merely beliefs (albeit implicit ones) that ensure the unity and consent of members of the scientific community, but as a powerful *means* of conceptualization and intervention in specific problem situations. Hence accepting a paradigm is more like acquiring and using many skills and abilities than gaining deep understanding and strong belief in the truth of any provisions.

> "No set of beliefs is sufficient to insure membership within a scientific community, which is best understood as a community of practitioners" (Rouse 1987, p.3)

The skills and dexterity that put up a paradigm contain the ability to apply learned concepts to a specific situation; the capability of application of mathematical tools (the faculty not only to solve equations but to choose the most suitable ones as well). One should apply them to a specific problem situation and be aware of the limits of their application, as well as of the possibility of bypassing them, etc.; then follows the application of corresponding measuring instruments and experimental techniques. It is important to emphasize that it is the method of *analogy* that is most often engaged in such a work, and not deduction, not logically correct conclusions of the particular principles from the general ones.

Note that scientists rather use paradigms to solve specific problems than belief in the absolute validity of their ontological, metaphysical components. The main thing is that they understand how to apply this knowledge as others do. At the same time, some scientists are completely ignored by their colleagues or even relentlessly excluded from the scientific community not at all because they share or do not share the beliefs of other researchers. They become sheer outcasts only when their work does not *interact constructively* with what others are doing. What really matters is (1) what is directly related to the work of other researchers and (2) the reliability of the results obtained. Hence the following belated Kuhn's remark is of special importance: "I later came to see my early emphasis upon perceptual Gestalts as misleading" (Kuhn 1993, p.16)

It is no coincidence that no set of beliefs is sufficient to insure membership within a scientific community, which is best understood as a 'community of practitioners.' What really matters is the ability of community members to recognize one another's work as satisfying norms of competent scientific practice, measured against the paradigmatic achievements that define the research field. Hence many stout clashes and misunderstanding might have been avoided had Kuhn ascribed paradigms not to communities, but to scientific *sub-cultures* sharing cultural forms, activities and aspirations.

Hence the fierce struggle between the old paradigm that is leaving the scene and the new one is not a struggle between the old and new 'Weltanschauungen,' but a struggle between the old and new *types of scientific life*. Diverse paradigms reorganize the world in various ways as a field of miscellaneous possibilities, offering differently organized challenges and opportunities for further fruitful research.

Therefore, as a rule, scientific revolutions consist not in changing worldviews, but in the emergence of new research instruments like centrifuges, supercomputers, and electron microscopes. For instance, it is this circumstance that explains the transition from classical cytology to modern cell biology (Bechtel 1993; Rheinberger 1997). Even the scientific revolutions which are associated with profound theoretical shifts are always connected with radical changes in instrumental and research practices. For example, as Andrew Pickering has ascertained, the revolution in elementary particle physics (that took place in the 1970s and 1980s and consisted in the advancement of gauge transformations) was engendered primarily by the emergence of novel, more powerful particle accelerators (Pickering 1984).

That's why in actual scientific life fierce controversies on the incommensurability of successive paradigms are rather rare and unique phenomena. Coming into existence of novel ways and methods of inquiring about Nature opens up such fascinating research prospects that it usually outweighs the quite commendable desire to comprehend which ontologies fit reality better. If one turns even to the science of the late XVIIIth century, s/he will find that the main drawback of phlogistonic chemistry was envisaged by Priestley and Lavoisier's contemporaries, not in the absence of empirical falsifiability of its concepts, but in its inability to direct further studies of new 'air' (i.e., gases) discovered in pneumatic chemistry.

As a result, Kuhn's concept of scientific revolutions reoriented philosophers, sociologists, and historians of science towards the scrutiny of manifold scientific practices, and not of scientific knowledge itself. Kuhn's legacy encourages one to apprehend scientific understanding, not just and not so much from the standpoint of gaining scientific knowledge.

Accordingly, science sets as its goal not so much the production of systems of grounded beliefs as the transformation of human abilities for practical and discursive interaction with the world. It is necessary to distinguish paradigms - beliefs from paradigms-achievements, which, as models, direct further research activities. As a result, scientific understanding is more about the practical ability to deal with a variety of relevant situations than the acceptance of stable, well-proven truths.

Incidentally, the neo-Marxist analogy with social revolutions is apparent. Like the social revolution, the scientific revolution consists in changing the *methods* of production. The criterion for the applicability and survival of a particular theoretical concept is grounded on research practice.

To sum up, as Hasok Cheng had nicely captured: the practice turn constituted an uprise against the long criticized yet always recurring misconception that science is an "ivory tower or a kind of *Glasperlenspiel*" empowered by its very distance from the affairs of ordinary humans. Consequently, diverse social and material practices are not to be taken as 'external forces 'influencing' the direction of research,' but as necessarily integral to knowledge construction

Eventually, Rouse's research project brought him to the following standpoint. Borrowing a model for understanding ecological niches from Richard Lewontin, he maintains that science is a niche phenomenon that should be comprehended as a form of co-production where multifarious scientific practices participate in constructing an environment. In his brilliant "*Social Practices as Biological Niche Construction*" he states that people's capacities, interests, and even lives are shaped by lifelong co-development with diverse environments. Forsooth, people are organisms whose bodily capacities, practical orientations, and continuing existence are constitutively entangled with the *biological* environments with which the human lineage coevolved and with which individual humans co-develop in diverse ways.

Eventually, the familiar differences between human ways of life and those of other organisms arose from co-evolution with the distinctive environments shaping human development in each generation. Ipso facto diverse yet i*nterdependen*t practices constitute the characteristic shape of human developmental and selective environments. These practices are shaped by the closely coupled *mutual interdependence* of those activities of others. These patterns of closely coupled, situated performance only exist through continuation and differential repetition of mutually supportive performances in sufficiently conductive circumstances. Those circumstances first and foremost embrace *other practices* that enable, support, or sometimes conflict with a given practice. Humans now sustain themselves as living bodies in a continuing lineage and develop in characteristically diverse ways by participating in *many interdependent practices.*

Finally one should speak about *complex networks* of interdependent practices, subpractices, and coupled performances making up those practices.

> "Humans have developed highly complex, multilevel ***interdependences*** among practices" (Rouse, 2014, p.18).

People's practice differentiated way of life is articulated horizontally and vertically and shaped by dynamic interactions among these forms of differentiation. Finally,

> "We gain a richer and more detailed grasp of scientific understanding and scientific practice by recognizing it to be an outgoing process of niche construction" (Rouse, chapter 9, p.287).

However, what are the inevitable *drawbacks* of such a patent startling approach? – In my view, the most apparent is coined by the following proverb of Martin Heidegger: "theoretical research is not without a praxis of its own" (quoted from Kochan 2017, p.67). Rouse's 'practical hermeneutics' (Kochan) takes science as a collection of interrelated interpretative practices, rather than as an abstract system of concepts and theories. These interpretative practices are exemplified by material activities in the scientific laboratory. Rouse accepts Heidegger's existential conception of science, indeed. However, he attempts to further elaborate it by exploring the ***constitutive role*** played by *material* practice in relation to concrete modes of scientific existence. Forsooth, for Rouse,

> "there is no theory proper apart from *applications* to concrete empirical situations, better put, the open-ended pattern exemplified by its canonical applications **is** theory…What the theory says about the world is not yet fully determinate but only emerges over time in the concrete uses of its concepts, uses which are embedded as such in *material* practices" (Rouse, 2013, p.9; my italics).

From his limited one-sided standpoint Rouse treats Heidegger as pan-theorist exaggerating the role of a theory in scientific research. In particular, he contends that the famous 'mathematical projection of nature,' which Martin Heidegger had located at the heart of modern science, is an inherently *theoretical* phenomenon. Rouse's 'Heidegger' defines science as a theory-driven mathematical projection of nature in which things show up as fully decontextualized and present-at-hand.

But Heidegger does not take the distinction between theory and practice as a fundamental one. Hence he does not identify the mathematical projection of nature with theory. In Heidegger's view, the mathematical projection provides the *existential condition* of possibility for ***both*** scientific theory and practice. The distinction between the two is therefore not fundamental, but instead derives from their shared existential basis in the abovementioned mathematical projection of nature. They are, so to speak, "two sides of the same existential coin" (Kochan).

We will argue, on the contrary, that Heidegger introduced his 'mathematical projection of nature,' as a substantially existential phenomenon which serves as a condition of possibility of *both* theory and practice in the sciences. Although scientific theory is necessarily enabled by practice, Heidegger resisted the urge to explain it reductively in terms exclusively of practice (Glazebrook 2000). On the existential account science is comprehended to be a mode of existence, a way of being-in-the-world, which brings to light things for *theoretical* understanding. Forsooth, the existentially decisive feature of science is, in Heidegger's view, neither empirical observation nor math modelling, but rather "the way in which Nature herself is mathematically projected" (Heidegger [1927], 2001a, pp.413-14).

Thus, to improve Rouse's pervasive account, one has to turn to a more detailed study of the genesis and advancement of phenomenology.

I.6. Husserlian and Heideggerian Abstruse Epistemologies: Eliciting the Hidden Facets of '*Structure*.'

I.6.1 A Brief Introduction.

As is well known, an American philosopher, historian and sociologist of science Thomas Samuel Kuhn, along with his pre-eminent adepts, stood at the origins of the so-called "Historical Turn" in the realm of philosophy of science, when the assiduous researchers increasingly began to realize that 'in order to comprehend what science is, one needs to understand how it advances.' The exponential growth of history-of-science trailblazing advances, obliged to this turn, overshadowed for the first time those epistemological ideas that stood behind them. It engendered the enticing impression that the works of this epoch of 'die Sturm und Drang' and, above all, thrilling "*Structure of Scientific Revolutions*," first published in 1962, constituted a *direct* generalization of history-of-science impelling data. There turned out an illusion that in order to do away with logical empiricism, it was enough to just punctiliously 'wipe your eyes' and look with a new, unbiased, fresh gaze at the history of science in order for novel epistemological discoveries to spill out of abundance.

Nevertheless, sober and deep elucidation about what significant epistemological results were *actually* obtained supplanted precocious enthusiasm. This, in particular, is convincingly evidenced by the collection of Thomas Kuhn's latest articles "*After the Structure of Scientific Revolutions*," already published after his untimely death

> "Now I think we overemphasized the empirical aspect of our enterprise (an evolutionary epistemology need not be a naturalized one)" (Kuhn, 2000, p. 95].

In this collection, one of Kuhn's goals was not only to indicate the significant changes in his epistemological position that took place after 1961, but also to reveal their *philosophical premises*. Nevertheless, true to his thrilling style, Kuhn rarely exposes names, focusing instead on the substantive side of the issue and positing his thought-provoking ideas as a direct generalization and a logical result of the work produced by a historian of science. However, his writings embrace rather important direct references not only to Charles Darwin and Immanuel Kant, but also to Ludwig Wittgenstein, Hilary Putnam and even Martin Heidegger.

In fact, Kuhn unfolds the startling analogy between the advancement of science and biological evolution as the basis of his concept. Uncoiling the analogy, Kuhn even somewhat departs from his distinctive tenet about the alternation of periods of authoritarian 'normal science' with involved periods of 'revolutionary development.' Now in his writings the periods of calm development within the framework of a single paradigm can be eventually replaced by the periods of revolutionary 'splitting' of a single research tradition into two competing traditions (first and foremost, the theory of relativity and quantum theory are meant).

It was the consistent drawing of this analogy that ultimately allowed Kuhn to unfold his philosophical position in the following way:

> "By now it may be clear that the position I'm developing is a sort of post-Darwinian Kantianism" (Kuhn 2000, p. 104).

Now a revolutionary change in science for Kuhn is no longer an evanescent 'mutation,' but rather a 'speciation'; the analogue of the biological population is, of course, represented by the habitual 'community of communicating specialists.'

On the other hand, a conscious and consistent obliteration of notorious correspondence theory of truth necessarily leads Kuhn to the conclusion that 'the world is somehow mind-dependent, perhaps an invention or construction of the creatures that inhabit' (Kuhn 2000, p. 101].

Of course, the basis of this design should constitute 'something stable, rigid and unchanging';

> "But, like Kant's *Ding an sich*, it is ineffable, indescribable, and stable. Located outside of space and time, this Kantian source of stability is the whole from which have been fabricated both creatures and their niches, both the 'internal' and the 'external' worlds" (Kuhn 2000, p. 104).

Thus, the ardent appeal to the philosophy of Immanuel Kant ('Ding an sich') and Hilary Putnam ('inner worlds') is apparent. Moreover, as Kuhn contends on the philosophy of the latter,

> "When he started talking about inner realism, I thought:' What a nightmare, now he speaks my language!'" (Kuhn 2000, p. 310).

Nevertheless, no less important for apprehending the origins of Kuhn's concept is, as I am going to demonstrate further, an appeal to Martin Heidegger's philosophy. Indeed, to begin with, in one of his writings, Kuhn explicitly uses such a pivotal concept of Heidegger's philosophy as '*being-in-the-world.*'

> "Some ways are better suited to some purposes, some to others. But none is to be accepted as true or rejected as false: none gives privileged access to a real, as against an invented, world. The ways of *being-in-the-world* which a lexicon provides are not candidates for true/false" (Kuhn 2000, p. 104).

But the abovementioned, I contend, cannot limit the active implementation of the Heideggerian distinct dictionary. Indeed.

(i) First, disclosing the content of the new paradigm created as a result of the scientific revolution, Kuhn portrays it as such when 'the individual parts suddenly come together in a new way'; however, the basic changes are not perceived gradually, step by step. They represent not only unexpected, but also "an integral transformation in which some part of the acquired experience is organized differently and reveals facts that were not noticed before" (Kuhn 2000, p.24). As we will elicit later, the holistic nature of the paradigm correlates perfectly with the holism of a 'style of thinking' (Heidegger) or a 'regime of truth' (Foucault).

(ii) As is well known, the Kuhn-Feyerabend startling tenet states that the new and old paradigms (Ptolemy and Copernicus, Aristotle and Galileo, Newton and Einstein) are *incommensurable*, since there is no neutral language for comparing them.

> "No other aspect of *Structure* has concerned me so deeply in the thirty years since the book was written, and I emerge from those years feeling more strongly than ever that incommensurability has to be essential component of any historical, developmental or evolutionary view of scientific knowledge" (Kuhn 2000, p. 91).

It turns out that the partisans of the new and old paradigms inhabit the different worlds, looking at the same things in different and even inimical ways. Further we shall expound that this position - i.e. the so-called 'Kuhn-Feyerabend thesis' - finds its direct correspondence in Heideggerian epistemology.

(iii) According to Kuhn, one of the most conspicuous examples of incommensurability thesis is the interrelations of Aristotelian and Newtonian mechanics. In Aristotle's physics the term 'motion' denotes a change in general, but not a change in the position of one physical body relative to another. For Aristotle, movement is a change of substance. In Newtonian mechanics, a moving body consists of particles of matter, and its qualities are a direct consequence of their interaction. At the same time, Aristotle's matter constitutes some substrate that is present where the body is. A specific body as a part, a piece of this substrate, differs from other pieces in that it imbibes a variety of qualities like a sponge. All the changes are connected with a change in qualities. Roughly speaking, qualities change (move), while the substrate itself remains unchanged. For instance, a stone, if not disturbed, always falls to the centre of the Earth. Fire rushes to heaven, bringing to fruition its natural properties in the same way as a corn realizes them in the process of its growth. And the fascinating example with the physics of Aristotle was actively implemented by Heidegger in writings devoted to the theory of knowledge.

(iv) The status of the new and old paradigms, as well as the status of the meanings of corresponding words, is conventional.

> "Each lexicon makes possible a corresponding form of life, within the framework of which it is possible to affirm and rationally justify the truth or falsity of statements, but the justification of the lexicons themselves can only be pragmatic" (Kuhn 2000, p. 249).

The aim of the present part is to reveal that the multilateral properties of the Kuhnian model (i) –(iv) are in fact not so much direct generalizations of history of science data relating to the first and second scientific revolutions, but rather constitute logical conclusions from the Heideggerian concept of truth and the growth of (scientific) knowledge. Furthermore, in my view the enrichment of the philosophy of scientific revolutions by accumulation and advancement of certain Heideggerian ideas would will contribute to a deeper understanding of the problem.

I.6.2. Husserl and Heidegger: Gnoseology vs. Ontology.

As is well known, the Heideggerian ontology sprung out of Husserlian phenomenology; it is the meticulous study of the latter that makes it possible to better comprehend the origin of the former. According to Husserl, phenomenology can be 'in the first approximation' elucidated as a continuous effort to survey our experience (and, accordingly, the 'things in themselves') without metaphysical and theoretical speculations. (Recall Husserl's famous appeal: 'back to things on their own!'). Husserl maintained that only 'suspending' or bracketing away a 'natural attitude' can ultimately turn philosophy into a definite and exact science. Accordingly, phenomenology is a science about consciousness, but not about empirically given things. It is no coincidence that the unrolling of Husserlian phenomenology commences with a sharp criticism of both 'psychologism' and 'naturalism.'

An inmost component of Husserl's concept dating back to Brentano (Husserl 2009) is the so-called '*intentionality*': the basic hallmark of consciousness consists in that it is substantially intentional, i.e. aimed at some of its own 'internal' experience. Brentano himself enunciated intentionality as the most vital trait of mental phenomena that distinguishes them from physical ones. Each belief, each desire has its own object: that which they believe in, that which they desire.

As a result, to highlight the structure of consciousness, one must distinguish between the acts of consciousness themselves and the phenomena at which they are directed. In particular, the knowledge of 'essences' becomes possible only when one 'brackets' all the premises about the existence of an external world (the so-called '*epoché*'). Hence, Husserl maintains that mental and spiritual processes have their own reality, independent of any physical basis. Cognition of this reality, he declares, constitutes the main goal of his epistemology. For her knowledge, he, just like Descartes, chooses the vantage standpoint of a systematic methodological doubt and develops a system of certain abovementioned procedures.

It is at this point that Husserl and Heidegger radically diverge. The latter argues that the most pivotal starting point is not epistemology, but ontology, the key to the construction of which is human existence or *Dasein.*

> "The fundamental ontology... must be sought in the existential analytics of Dasein" (Heidegger, [1927], 2001a, p. 13).

While Descartes and Husserl incessantly strived for an irrefutable foundation for philosophizing, Heidegger starts by questioning the very formulation of such a query. And immediately makes a reservation that the temptations to make this statement with the help of Hegelian dialectics should be intransigently avoided. With Hegel, dialectical development ends with the return of the Absolute Idea to itself. With Heidegger, this completion is quite impossible. According to Heidegger, historicity cannot be understood according to Hegel (and Marx) as a return from a*lienation* (Biemel 1987, p. 22).

Unlike Descartes and Husserl, a person for Heidegger is not an individual impudently snatched from the world, a kind of cognitive Robinson Crusoe who consciously establishes his dubious relationship with the world. Man is not just 'abandoned in the world.' He is already an inextricable part of *this* world. In principle, he is not able to detach himself from the world and confront it as an objective and impassive Cartesian 'cogito.' For Husserl, phenomenology is a means of getting rid of all theoretical prejudices and premises in order to break through to 'things on their own.' For Heidegger, the essence of phenomenology is "to let that which shows itself be seen from itself in the very way in which it shows itself from itself" (Heidegger [1927], 2001a, p. 34).

It is the phenomenology conceived in this way that allows one to outline the ways of answering the main problem of Heideggerian ontology – '*What is Being*?' - After all, "the being of everything is less capable of ever being something like this, "something else 'is behind it,' which does not appear"" (Heidegger [1927], 2001a, p. 35). Ontology is possible only through phenomenology: "everything again stands to get rid of the constructed concept of truth in the sense of correspondence" (Heidegger [1927], 2001a, p. 33).

In order to clarify the basic difference between Heidegger's phenomenology from Husserl's, one may turn to the following Husserlian faithful example (Feinberg, Leiss 2007). If we go around any material object - for instance, a table, then we will observe it from different sides. Each perception represents the same table, but each one is different from the other. Usually we make the 'apparent' assumption that our perceptions are held together by the 'hard fact' that they all refer to a *real* table that is disposed in the outside world. Husserl does not rebut it. But he contends that this assumption makes it impossible for us to 'perceive the actual process of organizing camera angles and holding them together in our minds.' In order to reveal the real cognitive process of perceiving an object through consciousness from its trenchant perspectives, one must prudently 'suspend the natural installation' of consciousness, redirecting attention to the so-called 'immanent structure of experience.' This is exactly what 'intentional correlation' consists in, the connection of acts of consciousness and their objects (table).

From the vantage standpoint of Husserlian phenomenology, 'acts of consciousness' engender 'the meaning of experience.' The multiple facets of a table, joined together, define what we call a 'table.' Masterly summarizing the above example, Husserl comes to the following paradoxical conclusion. Consciousness is commonly treated as something existing 'inside' the psyche. The latter is an object that is peremptorily enchained to another object - the body. According to this 'everyday' model, one commonly apprehends the encounter with the table as the interaction between the table itself and the light rays reflected from its surface and reaching the retina. However, Husserl contended that this blunt causal explanation "takes us away from experience as such." Therefore, it is necessary to promptly '*suspend the natural setting*' in relation to our consciousness and to our body.

One comes to the odd conclusion that Husserlian phenomenology does not actually give an answer to the thrilling question: "What is Being?" - Yet the latter is pivotal to Heidegger's philosophy. Consider, for instance, his fine master's thesis "*On the multiplicity of meanings of things according to Aristotle*," accomplished with the enunciation of the following startling question: "if being is being described as possessing *many* meanings, then which of them should be considered leading and basic?"

Husserlian enticing philosophy does not allow to answer the question. Husserl's being is defined as a random correlate of a whole series of meanings, each of which is isolated from the others by a corresponding act of intuitive comprehension of the essence in phenomenological analysis. As a result, every philosopher always conceives merely one aspect of being, and all the others are displaced, mercilessly 'suppressed' by him. How to avoid this, Heidegger inquires, how to impose the *integrity* of the comprehension of being?

In "*Being and Time*" he starts by stating the hard fact that 'the being of being is not existing.' So, from what being it is necessary to read out the subtle meaning of being, from what being should the 'disconnection' of being take its origin? - Heidegger maintains:

> "The contemplation of the existential question therefore means highlighting a certain being - the asker - in his being ... This being,

> which we ourselves always have and which, among other things, has the existential possibility of asking, we terminologically grasp as Dasein" (Heidegger [1927], 2001a, p. 7)

It is exactly a man - Dasein - who is the one who is able to perceive the being in its unconcealment mode. Heidegger points out that the meaning of the word 'being' is constituted by our comprehension, which is grounded in the temporal structure of our understanding. It is this temporality that amply furnishes the ground for further Heideggerian contemplation of the 'transcendental horizon' of being as such.

As a result, the Heideggerian ontology in the main does not strive to develop a global, comprehensive and common understanding of being for all times and peoples. It merely intends to give an account of the existential methods (existentials) - the fundamental structures of the questioner. These existentials should be distinguished from the definitions of the non-human being that Heidegger christened 'categorical definitions.'

In other words, one should try not to provide overall definitions of Being itself, but to contemplate Dasein itself where and when it is the closest and most widespread - in its *everyday life*. The essence of the phenomenological approach consists in the fact that we should not proceed from some ideal draft project, but should 'grasp' (halten) a person in his common ways of being. This will make it possible to underscore what underlies mass behavior, i.e. the '*Existentials*.'

According to Heidegger, when defining being, we must be fully aware of what concepts and from what time we are bypassed, instead of acting as if we are talking about some eternal collection of unchanging impeccable truths.

On the other hand, according to Husserl, if we want to know what phenomena really are, we should envisage consciousness 'in itself,' and not the results of empirical sciences. It is the subject that provides the conditions for all determinations of all objects of experience.

Heidegger admits with Husserl that the 'being' of all things consists in all the meanings that we receive in the process of understanding them. Nevertheless as the main shortcomings of Husserlian phenomenology (besides the fact that the latter does not give an answer to the question 'What is Being,' leaving us with a pile of chaotic set of meanings of

Being), Heidegger points to the following two interrelated traits (Frede 2006).

1) He strongly disagrees with classical consideration of the subject of knowledge as 'impersonal and transparent ego.'

2) Heidegger asks perplexedly how it is possible to 'bracket' the outside world.

As a result, Husserl's main 'fallacy' is that he magnifies the significance of the *theoretical approach* as such. Human activity as a whole is not determined by mere consciousness - let one recall the fine example of a man with a hammer. A man hammers nails into a board unconsciously, and he succeeds. He begins to wonder what the true meaning of the hammer is and whether he really needs to hammer nails only when he hits them on the fingers.

Correspondingly, according to Heidegger and his fervent disciples, the pivotal hallmark of our attitude to the world is not cognitive, but *practical.* It is the boundless expansion of practice that underlies the growth of our knowledge of the world. (Wrathall 2011). But this does not at all imply that one must now turn to the Marxist thesis about the 'revolutionary transformative role of practice.' Heidegger takes individual human existence to be primarily concerned with the attenuating prospect of his individual death, and this is stronger than all social considerations combined. The efficacious concept that *embraces* all aspects of being, which knocks these aspects into a single whole, is 'concern' (for more details, see Frede 2006, pp. 42–48). It is the daring enunciation of the structure of care that allows Heidegger to proclaim that our being is simultaneously 'being in the world' as an organic whole.

I.6.3. Heidegger on Truth and Delusion.

Thus, according to Heidegger, cognition constitutes 'the Dasein modus crucial for being-in-the-world.' But then 'What is the Truth?' – The paper with a self-explanatory title - "*On the Essence of Truth*" (2001b) - commences with genuinely Heideggerian analysis of expressions of common language. In what contexts in our daily practice, in ordinary life, do we most often use the exciting adjective '*true*'? For instance, on the one hand, we are talking about 'true gold,' meaning by the latter such metal, which corresponds to what we usually mean by 'gold.' But, on the other hand, we call true not only gold and other *things*, but also some *statements* about things. A statement is apparently true if what it means is consistent with what is actually going on.

The *dual* nature of this correspondence is reflected in the classical, traditional definition of truth: *Veritas est adaequatio rei et intellectus*. On the one hand, this proposition is translated as follows: truth is the correspondence [Angleichung] of matter to knowledge. However, it can be translated as follows: truth is the correspondence of knowledge to matter. And this is not a simple inversion of one sentence to another. In each case, 'intellectus' and 'res' are conceived differently. In order to understand why, one must turn to the time of creation of the above Latin phrase - i.e. to the Middle Ages.

And then it will become clear that the first meaning of the proposition *Veritas est adaequatio rei et intellectus* expresses the Christian theological belief that material things, 'matter' are created (ens creatum) only in accordance with the ideas presented in advance in *Intellectus Divinus*, i.e. in the Divine Mind. Yet 'intellectus humanis' is 'ens creatum' too. As the ability that God has endowed man, he must also be in line with his idea. And if all beings are diligently 'created,' the possibility of the truth of human knowledge is rooted in the fact that both matter and the utterance about it descend the level of an idea in the same way. Therefore, they fit together - on the firm basis of a single divine plan of creation!

If we do not accept the main tenets of Christian theology, then again we are left alone with the problem of substantiating notorious 'correspondence theory of truth.' We are again faced with the 'Cartesian problem': if the material and the ideal belong to different realms, how is certain and reliable knowledge of reality possible? What is the trenchant correspondence between the statement about the thing ('this coin is round') and the thing itself? The coin is made of metal. The statement is not material at all. What corresponds to what here?

Apparently, the correspondence in this felicitous example cannot be material in nature. What is maintained in a statement that represents is only that a thing exists. What is disclosed in this case was previously experienced in Western thinking as 'what is represented' and was peremptorily called 'being.' Therefore, the traditional, 'commonplace' attribution of truth to statements alone should be disavowed. The enticing possibility of 'standing in the light of being' as an internal condition for the possibility of correctness is rooted in freedom. 'The essence of truth is freedom.' Freedom is involvement in the disclosure of being as such.

No man possesses freedom as property. On the contrary - in the best case, freedom, existence, revealed Dasein possess a person. Only an existential man is historical. The nature does not have history.

> "Each kind of standing of a historical person in the light of being ... is rigged and introduced into existence as a whole by this fine tuning" Heidegger 2001b, p.307).

It goes without saying that this integrity is incalculable and incomprehensible from the point of view of everyday calculations and habitual worries. Absorbed in daily affairs and cares, a person exposes his being to oblivion, 'forgets his being as a whole.' Therefore,

> "he is all the more mistaken the more selectively he chooses himself as a subject, as a standard for all being" (Heidegger 2001b, p.309).

And, of course, the salutary remedy that helps a man to conceive his being as a whole is philosophy.

This aspect of Heidegger's epistemology can be better comprehended if one again turns to its substantial source - the philosophy of Edmund Husserl. It is better to begin exhibiting this source is with a trenchant example, often cited by Heidegger himself (Heidegger 1971, p.15). Consider a tree blossoming in a meadow that a person is looking at. At the same time, the natural sciences contend that we, in fact, do not envisage any tree, but in fact perceive a void in which the wave functions of electrons are bustling in different directions at great speeds.

Where is the tree located? Is the tree located in 'consciousness' or is it blossoming in a meadow? Does the meadow lie as an experience in the soul or is it spread on the ground? Is the earth in our head? Or are we standing on the earth? "We must go back to earth, says Martin Heidegger, and see a tree in a meadow, but not atoms, we must realize that we are dealing with *things themselves*, and not with their ideas in consciousness." Who gave science the ultimate right to call such thinking blunt, inferior and pre-scientific, exposing itself as the only measure of thinking?

From here comes Husserl's appeal 'to return to the things themselves!,' creating a philosophy of 'direct experience.' His faithful student Martin Heidegger applied Husserl's phenomenological method to unfold the meaning of not only ordinary perception, but also human existence as well.

> "With the leading question about the meaning of being, search stands with the basic question of philosophy in general. The way of working out this issue is phenomenological" (Heidegger [1927], 2001a, p.27).

But if Husserl in his phenomenology relies on the language of consciousness, then for Heidegger the subject of experience is not Husserl's 'pure consciousness,' but a really existing individual, 'a holistic acting I,' which is substantially involved in the realm of objects with which it interacts. Habitual 'meanings' sprung out of precisely this interaction.

Now - with Heidegger – 'intentional correlation' unites human being and the world in such indispensable unity, which Heidegger called "*being-in-the-world*." Furthermore, the Heideggerian 'world' does not mean Nature, but excellently correlates with common expressions 'world of theater,' 'world of cinema,' 'Eastern world,' etc. There can be quite a lot of similar worlds. None of them is certainly 'subjective,' but equally none can be unique and absolute. Ultimately, each of these worlds constitutes a Weberian semantic context of actions, but by no means the unpretentious sum of existing things. 'World' in the above sense cannot be conceived without proper correlation with the acting subject, 'immersed' in this world.

As a matter of fact, the label 'phenomenology' obliquely expresses the peremptory maxim 'to things themselves!.' Therefore the concept of phenomenology is amply established through the characteristic of what is meant by both components of the term, '*phenomenon*' and '*logos*.' The first term, a 'phenomenon,' is exhausted by indicating that a phenomenon constitutes a letting you know about yourself through something that 'reveals itself to itself.' Sooth to say, "the being of everything is less capable of ever being something like that, "behind which" is something else, 'which does not appear'" (Heidegger [1927], 2001a, p. 35). But the second term, λόγός, is 'translated,' i.e. is ascertained not only as 'reason, judgment, concept, definition,' but also as '*foundation* and *relation*.'

> "And only because the function of the 'logos' ... lies in the revealing ability to see something, the 'logos' can possess the structural form of 'synthesis.' Synthesis here does not mean the connection and cohesion of representations, the manipulation of mental processes, regarding the interrelations of which there should then arise a 'problem': how they as an internal are consistent with the physical outside. 'Sin' here has a purely apophantic meaning and implies: to see something in its compatibility with something, something as something. And again, since λόγός is a gift to see, it can be true or false. Everything again stands to get rid of the constructed concept of truth in the sense of 'correspondence.' This idea is in no way primary in the concept of αλήθεια ["aletheia"]. The 'truth' of the logos as truthness, αλήθενειν, implies: to remove the being about which it is in λεγειν as ... from its concealment and to let it be seen as unconcealed, to reveal it" (Heidegger [1927], 2001a, p. 34).

As a result, according to Heidegger, a man is one who is able to perceive the existent in its unconcealment, and, therefore, concentrate at the same time - λεγέιν - on its strong connection. Λεγέιν means 'reading in the sense of reckoning – collecting.' The Logos reveals and makes apparent this strong connectivity. In "*Being and Time*," Heidegger contrasts the truth as 'conformity-reconciliation' to the truth as 'non-concealment' ('aletheia'). In this sense, the incessant attitude of constant *unconcealment*, '*openness*' can be taken as *the essence of truth.*

For Heidegger's distinctive answer to the question 'What is Truth?' the following superlative example gives a lot. If I hammer a nail into a board with a hammer, then I unfold the way of proper using a hammer. In this broad sense, my action with respect to the hammer is true - in using the hammer I 'bring to light the Divine' what the hammer is and how it is felicitously applied. Or, for example, if a certain building, like a medieval cathedral, supports believers in their life in this ramshackle world, revealed to us by the grace of God, the cathedral is also true in a broad ontological sense.It functions by stubbornly bringing to light the divine that is essential and especially important in this world, and by supporting the revealing practices of the people who make their abode in it.

Truth, according to Heidegger, is 'uncoveredness' = unconcealment = Unverborgenheit = aletheia = Wahrheit. Accordingly, the history of human cognition is 'a series of epochs of unconcealment,' an alternation of multifarious felicitous world disclosures. The essence of such a history shows itself in "the separation of the truth of entities from possibilities of essence that are kept in store and permitted, but in each case not to implemented" (Heidegger; cited in Wrathall 2011, p.7). What separates one historical epoch from another is that each has its own thriving style of 'productive seeing.' As a result, the unconcealment of beings is the 'anticipatory gathering' that lays out certain properties and relationships as salient. Therefore, '*essences are historical*': they show up differently as dispositions for the world change.

Of particular note is Heidegger's notorious skepticism towards the so-called '*propositional truths*' or truths of utterances. Why, Heidegger asked, should we limit our contemplation of truth to only the consideration of the truth of sentences, i.e. only to 'propositional truth'? - Let's turn to the ordinary language: 'Vovan and Dimon are true friends,' 'this armoured- jacket is truly magnificent,' etc. In addition, recall the Gospel of John. Jesus Christ proclaimed: 'I am the way, the truth, and the life!' Heidegger punctuated that truth in its original sense does not refer to utterances or even objects, but to Dasein itself. Only in the sense of realization of unconcealment can certain 'objects' vigorously be called true. 'Truth' and 'falsehood' are indispensable characteristics of human activity in terms of concealment of faces behind masks and deception. For the most part, Dasein does not reveal itself and remains hidden, which is characteristic of idle, common babbling (Gerede).

Concealment reigns in such a non-assertive interaction with the world, in such pre-predictive comportment when the world is experienced as devoid of trenchant certainty, devoid of propositional expressiveness.

Heidegger contends that in our daily experience we perceive the world precisely as propositional concealment. In our predicative experience of the world, things are conceived by us in terms of our practical interaction with them. Things defined in a practical way are immersed in a complex variety of interactions with other objects, practices, and goals.

But in a logical judgment, on the contrary, our experience is subject to considerable confinement. We dim the whole context that spreads before us in order to concentrate on one particular feature of the situation. We are tearing some peculiarity out of its 'unarticulated totality.' Therefore, logical judgment expresses the state of affairs in a different way than they were bestowed to us in natural perception. The properties of things that were transferred to them had been indeed previously possessed by things, but now, with the help of assertive judgment, they are torn out of their inconceivable context and put on merciless public display.

In the 1929-1930 lecture course famously devoted to the "*Fundamental Concepts of Metaphysics*," Heidegger contends that the world should be comprehended as the predominance of the 'pre-logical bulge' of things 'as such and in general' (Heidegger 1995). Heidegger ardently retorted such accounts of comprehending the essence, which determine what a thing is, with the help of some inherent property that all things of a certain class should possess. On the contrary, he maintained that the essence of a thing is determined by that in the light of which this thing is brought into a state of unconcealment. The essence of a thing is determined not by the property that the thing currently possesses, and not by the abstract type that it embodies, but by the one in the light of which we accept it as it is.

It is such a contemplation of the essence - as such a setting that allows us to contemplate certain properties of things as defining, leading, basic - to equivocate the difficulties of essentialism and realism. As an example of the *historicity* of the essence of things in different milestones, Heidegger proposes to compare two different understandings of the essence of gold: (1) medieval (proximity to God) and (2) modern (ability to transform into a valuable resource). In each of these stages, the understanding of the existence of things is hidden when we cannot operate it in our experience of things in the world.

And so is a man: the fact that we are in the truth is indispensable part of our human essence. Essences are fatally historical, and this proposition refers to the human essence too. What it means 'to be human,' or, to put it more Heidegger-style, something in the light of which something reveals itself as human, changes in history. This historicity of human nature is associated with a change in truth and in unconcealment, because the way in which we comprehend ourselves is rooted in the way in which the world exposes itself.

But the recognition of the existence of various stages of openness or 'regimes of truth' (Foucault) inevitably raises the question: can these stages be properly compared with each other and what are the lucid criteria according to which we distinguish a more uncovered era from a less uncovered one? - To answer these questions, Heidegger refers to the origins of ancient philosophy - to Plato, to his *ne plus ultra* 'Myth of the Cave,' portraying the prisoners enchained to one of its walls, able to contemplate only the shadows of things, but not the things themselves that disgusting slaves pass by the entrance to the cave (Wrathall 2011).

Plato strives to compare different types of unconcealment and to choose one that is more unconcealed than the other. In fact, the 'shadows' on the wall of the cave, i.e. everyday objects and situations that we constantly encounter in our lives are also unconcealed (i.e., available for appropriate actions, for appropriate comportment in relation to them). What then allows us to maintain that the objects and situations that appear to us in the light of the Platonic dazzling 'theory of ideas' are more unconcealed? - Apparently, Plato uses an implicitly similar criterion: the highest form of unconcealment constitutes the form that makes the lower form possible.

Plato's telling argumentation is based on the assumption of the primacy of ideas and their knowledge in relation to other practices or types of knowledge of the world, i.e. on the role that cognition of ideas plays in allowing more *practical* forms of comportment. Due to the fact that the prisoners do not yet have practical skills in working with ideas, they will be confused, embarrassed by objects articulated in the form of ideas. Therefore, the release of prisoners from captivity and their falling into the 'real world' will at least initially confuse them: presenting things to prisoners in a new light will not be associated with new practices necessary to interact with things. Until the prisoners plunge into the world of new practices and acquire the habits necessary for communicating with new things, they will not be able to free themselves from everyday situations.

As a result, the history of being falls, according to Heidegger, into four distinct periods.

> (i) The Greek period in which being was primarily understood as *physis* or as a self-arising nature.
>
> (ii) The Middle Ages, when being was primarily understood as *divine* creation.
>
> (iii) The Modern period, where beings became objects that could be controlled and penetrated by *calculation*.
>
> (iv) The Technological period, in which what it is comprehended as standing *reserve*, that is, as being constantly available for flexible reconfiguration, and thus maximally exploitable.

Technique is a type of revealing concealment. The essence of technology is the removal from concealment, i.e. realization of truth.

> "The hydroelectric power plant is not built into the river in the same way that the old wooden bridge that has been connecting one bank to the other for centuries is built into the river. Rather, the river is built into a hydroelectric dam" (Heidegger [1954], p.80).

The removal from the concealment that modern technology has captured is in the nature of representation in the sense of extractive production. Extraction, processing, accumulation, distribution, transformation are types of removal from the concealment. And if, searching and contemplating, a person begins to explore nature as a certain area of his imagination, then it means that he is already captured by that type of discovery of concealment that forces him to step on nature as the object of study standing before him.

Heidegger calls the exciting challenge that focuses a person on putting everything that comes out of concealment as 'der Gestellt.'

> "We call der Gestellt the collecting beginning of the installation… that method of unconcealment that governs the essence of modern technology, without itself being anything technical." (Heidegger [1954], p. 87).

The physical theory of Nature of modern times prepared the way, first of all, not for technology, but for the essence of modern technology. Physics of the New Age is an early messenger of der Gestellt not yet known in its origins. According to the historiographical account of time, the beginning of modern natural science falls on the XVIIth century. Machinery, on the contrary, develops only from the second half of the XVIIIth century. But a later event for historical factography - modern technology – 'in its ruling essence is an earlier event.'

Physics will never be able to refuse the following: the fact that Nature makes itself felt in some calculated, fixed way, allowing itself to be represented as a system of informative data. Precisely insofar as the essence of modern technology is rooted in der Gestellt, it is forced to apply exact ('mathematized') natural science. And indeed, der Gestellt is the collecting principle of that establishment that puts a person in the revelation of reality by the way of placing him as existing. By relentlessly drawing a person into the supplying production, der Gestellt sends him on a certain path of revealing concealment.

Heidegger's ardent discussion of the intricate history of truth explores the history of different understandings of what entities truly are. It amounts to thinking through historically different ways of determining what is steady and stable in the entities one encounters. An entity is 'uncovered' when it is available to be readily taken up into our impelling activities, when it amply *meshes with common practices*. In Heidegger's daring vernacular, 'the essence of truth' is 'the truth of essence.' Respectively, a metaphysic 'is the truth of entities as such and as a whole,' while the history of metaphysics is the history of diverse 'epochs' of truth, that is, various unified understandings of what entities truly are, and correspondingly multifarious views of what the privileged attitudes are for disclosing and grasping the ultimate truth.

I.6.4. Heideggerian Epistemology.

Turning to Husserlian phenomenology allows us to better comprehend the correlation of ancient and modern science, which is of particular interest to us in this work. According to Husserl, science commences with what we apply to the 'life world' (lebenswelt) - the world that is given to us as real in our concrete common life - the 'garment of ideas' that suits it perfectly. It is thanks to the latter that we take for 'true being' that which is merely a method intended to improve crude foresight in infinite progress. Accordingly, the 'prepredicative' world, the ontic universe can be 'thematized' in several possible ways. Therefore, as a result of a truly transcendental 'epoche' and transcendental reduction, we are convinced that

> "This world has not disappeared, only in the course of an epoche carried out in consecutive order it is taken into account purely as correlate of subjectivity, which bestows it with an existential meaning, subjectivity, by virtue of the effectiveness of which it generally is" (Husserl 2007, p.205].

Accordingly, our theme, the result of the thematising process should not be

> "the world as such, but exclusively the world as constantly given to us in the changing ways of its givenness" (Husserl 2007, p. 208).

Therefore, - continues the thought of Husserl Heidegger, - using the word 'science' today, we mean by this word something other than the 'scientia' of the Middle Ages or επιστήμή of the Greeks. Greek science was never exact and did not need exactness: it makes no sense to say that modern science is more exact than ancient.

> "It is also incorrect to say that the Galilean doctrine of the free fall of bodies is true, and Aristotle's doctrine of the desire of light bodies upward is false; for the Greek perception of the essence of the body, the place and the relationship of both rests on a different interpretation of the truth of existence and, accordingly,

> determines a different way of seeing and studying natural processes. It would never occur to anyone to argue that Shakespearean poetry went further than Aeschylus one. But it is even more inconceivable to say that the new European perception of being is more true than the Greek one" (Heidegger 1972, p. 59).

The main process of the new time is the conquest of the world as a picture. The transformation of the world into a picture is the same process as the transformation of a person inside of being into a subiectum. In this regard, humanism denotes how things in general are interpreted and evaluated from person and according to person. Only where the being has become the subject of presentation, does the being in a known way lose being. Representation is no longer revealing oneself to things, but grasping and comprehending. Therefore, the modern European interpretation of being is far removed from the Greek. A Greek man exists as accepting being, therefore for the Greeks the world cannot become a picture.

> "No subjectivity in Greek sophistry is possible because a person here can never be a subject; he cannot become one because being here is Dasein, and truth is unconcealment" (Heidegger 1972, p. 81).

In order to fix the difference between modern science and Greek, we must remember Heidegger's dictum that modern science is a substantially "*mathematical project of nature*" (Glazebrook 2000). The essence of what is called modern science is in research; modern science needs in advance an 'open sphere' for its deployment. This sphere is revealed due to the fact that in a certain area of existence (in nature) a certain 'comprehensive scheme' of natural phenomena is outlined. This sketch is

> "a project of what nature should continue to be in front of the sought after knowledge of nature: a closed system of moving point masses oriented in space and time ... Inside this general scheme of nature, any natural process should find its place. The natural process enters the field of view as such only in the horizon of the general scheme" (Heidegger 1972, p. 60).

It can also be stated that

> "the natural-scientific method of representation explores nature as a calculable system of forces [...] Only because modern physics is fundamentally *mathematical* can it become *experimental*" (Heidegger 1978, p.318-19).

The hallmark of the experiment is that it begins with the basis of a certain law. To set up an experiment means to imagine a condition under which a certain system of motion can be traced to the necessity of the processes occurring in it, i.e. make amenable in advance for calculation. Modern research experiment is not just an observation, more precise in level and scope, but a completely different kind of method of confirming a law *within the framework* and *in the service* of a certain project of nature.

What perception of existing and what understanding of the truth engender the transformation of science into research? - First and foremost, in science only what becomes an 'object' is taken for existing. This objectification of the existing takes place in the 'presentation,' with the goal of setting all things in front of themselves so that the calculating person can provide for himself on the part of this existing.

> "A picture of the world, essentially understood, means, therefore, not a picture depicting the world, but a world understood in the sense of such a picture" (Heidegger 1972, p. 69].

Therefore, to elicit this facet of Heideggerian epistemology one has to take a closer look at Heidegger's concept of 'mathematical project of nature.'

I.6.5. Heidegger on Modern Science as Mathesis[1].

Heideggerian conception of the 'mathematical project of nature' is organically grounded on the Greek concept of 'mathesis' or the 'mathematical' (das Mathematische). For Heidegger, 'mathesis' refers to that conspicuous characteristic of modern science which vigorously distinguishes it from both ancient and medieval science. Viz., it is not reliance on facts, as such, which is decisive for modern science, but 'the way the facts are concerned.' Thus, it is not that facts, experiments,

[1] The content of this unit relating to Heidegger's view on mathesis is a revised and condensed summary of Jeff Kochan's 2017 book "Science as Social Existence."

calculation and measurement are used, but *how* and to *what end* they are deployed, which distinguishes modern science.

It should be pointed out that Heidegger distinguishes the following two fundamental features of modern science as mathesis.

(1) '*Work experiences*,' or "the direction and way of controlling and using what is."

(2) '*Metaphysics*,' or "the projection of the fundamental knowledge of being, upon which what is establishes itself as knowable."

Wherein the experimental set-up is a material instantiation of, as well as the material basis for, the metaphysical projection. Therefore Heidegger's account of modern science in mathesis terms recognizes a central place for the *material practices of experimental science,* - but without an important role of metaphysics.

As a scientific practice, mathesis is ruled by certain norms. In this connection, the image of the demiurge reflects the impulse of some early Greek thinkers to construe thinking and things in terms of craft production.

> "All forming of shaped products is affected by using an image, in the sense of a model, as guide and standard" (Heidegger 1967).

Thus, the normative aspect of mathesis, as a productive practice, stands as a priori image in the experience of working with things, as a projective image guiding the pursuit of a specific and determinate knowledge of those things. This image (Grundriss) constitutes the basic blueprint. And on the one hand, the blueprint helps to guide the act of production. Yet on the other hand, it represents, schematically, the end result of production. As a distinctive model of scientific practice, it captures both the *how* and *what* of that practice.

According to the author of "*Sein und Zeit*," it is the kind of end-directedness possessed by modern scientific things which distinguishes them from their ancient and modern predecessors, namely, their directedness towards a basic *blueprint*, a single regulative ground plan, in respect of which they are let to be what they are. Yet the author of "*Being and Time*" stresses again and again that the essence of scientific thing is not something one receives from it, but something it possesses as a result of the *socio-historically conditioned metaphysical projection* within which it is let be what it is. Not even an infinitely powerful God could grasp the intrinsic, independently existing essence of a thing, because *no such essence exists*.

All the abovementioned helps to elicit Heidegger's account of the essential features of the first scientific revolution that until recent times was a focus of profound clashes between the philosophers of science. For instance, for Alexandre Koyrè, the revolution's hallmark- mathematization – marks a *radical* historical discontinuity in knowledge-making practices, and, more specifically, a sudden and profound usurpation of Aristotelian natural philosophy by a resurgent Platonism.

However, the startling idea that early-modern science was inaugurated by a sudden, radical, and complete-in short, a revolutionary-epistemic break with the past has been largely abandoned by modern historians of science (see the further units).

Thus, Heidegger's account of modern science as mathesis had commenced with his insistence that facts, measurement, and experiment, broadly construed, figure as continuous threads running from modern science all the way back through medieval to ancient science. Moreover, the first scientific revolution was sparked not by decisive *rejection* of tradition, but instead by the *consolidation* of an existential possibility which had, until that time, lain relatively dormant within that tradition. In this way, Heidegger's distinctive use of the term 'revolution' [Umwälzung] harkens back to an older meaning of that term. It is not a sudden and radical break with the past, but instead a transformative, as opposed to atavistic, *return* to it.

At its root, this mathematization had to do, not with numerical practice merely, but with the development of a *uniform measure*, a basic blueprint, circumscribing the 'thingness of scientific things.' It was the mathematization of final (Aristotelian) causes, rather than their rejection, which allowed physical apparatus to flood into the knowledge-making practices of early-modern natural philosophy. Heidegger grounds his account on mathematization of the pre-modern Aristotelian notion of a final cause in our experience of the thing as ready-to-hand within a work-world. That is, as something with which we are at work. According to him, such a thing has an end – directedness. When one works with (rather than against) it, he lets it be what it is in its directedness towards some end. He lets it, so far as is possible, fulfill that end. This end is the final cause of the thing, that for the sake of which one lets the thing, in the course of working with it, be what it is. Note that Heidegger argues that, in the case of early-modern scientific work, the end of the thing is a ground plan, a *basic blueprint*. As the final cause of the thing, the basic blueprint provides the measure against which our work with the thing makes sense.

Eventually, all potential scientific things will be necessarily circumscribed within the experimental realm laid out according to this measure. This circumscription forms the true basis of mathematization. Through mathesis, our knowledge may then be developed into something more specific and determinate, and this determination may take a number of different shapes.

Mathesis, as the mathematical projection of the thingness of things, as a basic projective structure in the subjectivity of the subject, served to facilitate the powerful influx of *both* numerical and instrumental practices into early-modern natural philosophy. Herewith scientific things came to be increasingly taken as a legitimate subject matter for artful manipulations.

Heidegger's account of mathesis is not so much a rejection as it is a resolution of the historical difference between what Thomas Kuhn later identified as mathematical and experimental traditions in science. Each tradition represents its own way of specifying and determining the same general and indeterminate knowledge of things. Kuhn's abovementioned traditions may thus be better described as sub-traditions within a broader tradition. The shared root of these two traditions in mathematical projection furthermore allows us to explain why these two traditions could eventually *come together*, especially, as Kuhn indicates (1977), in the case of XIX-th century physics.

Note that Heidegger defined mathesis in terms of a twofold reciprocal relation between a mathematical projection of nature, on the one hand, and work experiences, on the other. Hence Heidegger's account betokens the way in which material practice and metaphysical project came to *mutually reinforce* one another, each rendering strength to the other, until a powerful new way of understanding and intervening in natural processes emerged, a novel intellectual and material culture, a new way of being-in-the-world. Thus, Heidegger was a reductionist champion of neither theory nor practice, but was instead engaged in gaining a better comprehension of the complex relationship between the two. Yet when it comes to history-of-science analysis, Heidegger puts his emphasis on one side of this dynamic, giving more attention to *intellectual* rather than to pragmatic factors.

Let us recall Newton's First Law of Motion, his celebrated principle of inertia.

> "Every body [corpus omne] continues in its state of rest, or uniform motion in a straight line, unless it is compelled to change that state by a force, impressed on it."

The phrase 'corpus omne,' or 'every body,' indicates Newton's rejection of the fundamental Aristotelian-Ptolemy distinction between terrestrial and celestial bodies. Furthermore, according to Heidegger, Newton's famous law is a formal articulation of an understanding of thingness which had already been expressed in Galileo's early account of free fall. Galileo's key innovation was to explain the difference in the time it takes for distinct bodies to fall to earth "not from the different inner natures of the bodies or from their own corresponding relation to their particular place," but from the external forces acting on them, for instance, the resistance of the air (Heidegger 1967, p.90).

Heidegger maintains that

> "it is precisely the projecting-open of nature in the mathematical sense that is the presupposition for the necessity and possibility of 'experiment'" (Heidegger [1989], 1999, p.113).

According to Galileo and Newton, the Earth is accelerating the thing out of its indigenous state of uniform rectilinear motion (or rest). In doing so, it allows the thing to become an object of thorough experimental observation. The point is that the thing, as mathematically projected, allows the experimentalist to ascertain the corresponding causal influences as *irrelevant* to the effect under study, in the interests of isolating just those cases responsible for that effect.

Respectively, to achieve the same observed effect in the behaviour of the two bodies is to have successfully eliminated the differences in their respectable causal nexuses. (Just as was demonstrated during the famous TV show on the Moon's surface). Thus, the experimental practitioner isolates the cause of a particular behaviour, or effect, and then demonstrates the regularity of this causal relation.

In principle, all observable effects might be thoroughly eliminated from an experimental system, thus leaving the thing entirely to itself, in a purified state of mathematically projected thingness. Just to quote Aristotle's "Physics": "art partly completes what Nature cannot bring to a finish." On Heidegger's account, the experiment does not control the thing, but rather the causal nexus in which the thing is situated, and it does this in the interest of letting the thing be what it is. The thing is 'released' [entschränkt] by natural science from muddling confines of everyday situation. Thus the mathematical projection is a necessary condition of possibility for experiment.

Heidegger insisted on the unity of things and thinking and argued that rule-following is not individualistic, but presupposes one's participation in a tradition, a kind of awareness of one's historical being-with-others. Hence Joseph Rouse's reproach in pan-theorism is too hasty. Indeed, according to Rouse Heidegger's mathematical projection of nature is a substantially *theoretical* phenomenon. In particular, Heidegger's emphasis on this allegedly theoretical projection allegedly leads him to dismiss experimental practice as merely incidental to scientific activity. Yet Jeff Kochan stresses that Heidegger's 'math projection' is the *existential condition* which resolutely makes modern scientific activity – ***both*** theoretical and experimental – possible.

It is this hallmark of Heidegger's standpoint that provides the following feature of modern science.

> "the modern physical theory of nature *prepares the way* [...] for the *essence of technology*," and it does so because it "pursues and entraps [stellt] nature as a calculable coherence of forces" (Heidegger [1954], 1977, p.25).

Just recall Bruno Latour's apt remark that "modern science transforms society into a vast laboratory." Or, as Heidegger himself had put it clearly,

> "modern science as a theory in the sense of consideration is a terribly decisive processing of reality" (Heidegger 1954, s.65).

Science establishes the real, i.e. she strives to ensure that the real always appears as the result of one or another *action,* i.e. in the form of foreseeable consequences of the *reasons* given under it.

Heidegger made it clear that 'stellen,' in his distinctive usage, reflects two different meanings: (1) "an ordering which challenges forth" [das herausfordernde Bestellen]; (2) a "producing which brings forth" [des hervorbringende Her-stellen]. Yet he makes a reservation that these two meanings are "fundamentally different, and yet they remain *related* in their essence."

To more succinctly express the abovementioned difference, Heidegger proffers the word Ge-stell or 'enframing' for the first meaning. And for the second meaning – 'poiesis,' which is meant to signify the kind of the non-modern technical production typical of art, poetry, and handicraft. Yet 'enframing' and 'poiesis' are essentially related since "both are ways of revealing." In good sooth, if placing a thing in the light of understanding means putting it in its proper place, the place where it may be encountered as what it is, then 'enframing' and 'poiesis' each suggest *fundamentally different* conceptions of a thing's proper place, and hence of its whatness.

Eventually, we need the concept of a 'final cause' in order to make sense of the artful manipulation of things. The final cause properly explains the *order* and *meaning* of such manipulations, grounds their intelligibility, thereby distinguishing them from random, unstructured behavior. It thus plays an important role in any comprehensive account of practice. It properly explains how a practice realizes its completion in a finished thing, how the practice lets the thing become what it is. In Aristotelian solid terms, to let the thing become what it is means to guide it to its proper place in the cosmos. A practice may thus be contemplated as an act of placing, regulated by a final cause. And there will be as many final causes as there are proper places in the cosmos.

As for framing, Heidegger's idea seems to have been that enframing challenges or orders things – forcibly places, or frames, them – into a single, homogeneous picture. Homogeneous unfolding allows for a productive co-understanding of nature among members of the scientific community.

According to Heidegger, the final causes guiding and regulating practice ultimately bottom out in the *existential possibilities of the subject*, possibilities which comprise its historical tradition. Hence for both Heidegger and the partisans of SSK, cognitive order is grounded in social order. And the latter is, in turn, produced in micro-social interactions between mutually receptive subjects who both compel and defer to one another in (and often by manipulating bits of) a common material environment.

The norms governing scientific behavior are not, finally, to be located in a system, of which each person is a constitutive part, but in the often ***mundane*** interactions between persons who, to a greater extent or lesser extent, share a history and a culture.

I.6.6. Interim Conclusions.

Thus, there are ponderable textual reasons to assert that there are correlations between the main provisions of the Kuhnian and Heideggerian ideas concerning the growth of scientific knowledge. What is the cause-effect relationship here, and is the content of the Kuhnian model of the structure of scientific revolutions a relentless consequence of half- conscious acceptance of Heideggerian epistemology? - The answer to this question is apparently beyond the humble scope of present book. The most likely assumption consists in that one can only talk about *indirect* influence.

For instance, via the distinctive writings of Alexander Koyré, in one of which, "*Etudes galileennes*" (1939) Kuhn, by his own admission, "just fell in love" (Kuhn 2000, p. 384). Note that Alexandr Vladimirovich (Volfofich) Koyrè, a French philosopher of Russian origin, studied at Gottingen (1908-1911) under Edmund Husserl and David Hilbert. Following Husserl's "*Cartesian Meditations*," he gave a series of lectures in Paris in 1929 and met with Husserl repeatedly. Moreover, following Edmund Husserl's later studies, Koyrè (1957, 1961) maintained that modern science has succeeded in overcoming the split, inherent in

Aristotelian science, between Earth and Space, since these were both taken as governed by the same laws. One of the chapters of his startling book on the Ptolemy-Copernicus breakthrough begins with stating that

> "it would be of inestimable value for the history and *phenomenology* of human thought were it possible to reconstruct step by step the development of the Copernican mind" (Koyrè 1973, p. 18).

Koyrè's philosophical standpoint is confirmed by another significant opinion:

> "Knowledge, reason, rationality, the possibility of elaborating a history of rationality [...] I would say here again we come across phenomenology, in someone like Koyrè, a historian of science [...] [who] developed a historical analysis of the forms of rationality and knowledge in a phenomenological perspective" (Foucault 1988, p.23).

In 1922 Koyrè started teaching in Paris at the 'Ecole pratique des hautes etudes' and became a close colleague of another Russian émigré - Alexandre Kojève, who eventually replaced him as lecturer on Hegel. It is well-known that Kojéve presented in his thrilling 1933-1939 Parisian lectures a subtle comprehension of the Hegelian "*Phenomenology of the spirit*" surveyed through a cognitive lense of Heideggerian ontology (Kojève 1980). A 1931 French translation of Heidegger's inaugural lecture '*What is Metaphysics*?' included a telling introduction by Koyrè, in which he describes Heidegger as "one of those great metaphysical geniuses whose influence marks an entire period" (Alexanre Koyrè, 1931, L'introduction des "Qu'est-ce que la metaphysique?," Bifur 8, 5-8 (5); quoted from Kochan, 2017, p.262).

All in all, the main result of the above parts consists in that the future development of philosophy of scientific revolutions should be grounded on the certain *compromise* between cognitive and social factors with the latter consisting in a kind of medley of Heidegger's epistemology and the SSK. Let us sum up the system of the obtained results first.

I.7. Summing Up: Scientific Revolution as a Social in the Broadest Sense Process.

Modern *social in the broadest sense* prerequisites of reconstructing a *scientific revolution* can be reduced to the following ones.

(1) The basic unit in reconstructing mature scientific change is the so-called '*practice of science*.' And the latter, both conceptually and instrumentally, is full of genuine social assumptions. Crucial to the practice of science is the idea that science is grounded on the '*public faith*' in it (Shapin). This is why it is important to keep constantly explaining *how* reliable knowledge is produced and *who* takes part in the process.

(2) Modern historians of science (see Shapin 1996 for review) now reject even the notion that there was any single coherent cultural entity called 'science.' There was, rather, a diverse array of practices aimed at understanding the natural world, each with variegated characteristics and each experiencing different modes and tempo of change. Furthermore, modern historians are now much more dubious of claims that there is anything like a '*scientific method*'- a coherent, universal, and efficacious set of procedures for producing scientific knowledge.

(3) Moreover, historians and philosophers of science take for granted now that science is a *historically* situated and ultimately *social* activity; so, it is to be comprehended in relation to the plenty of diverse *contexts* in which it occurs. They are more and more dissatisfied with the common manner of treating scientific ideas as if they floated freely in conceptual space. Thence they now wish to comprehend the concrete scientific practices by which ideas are produced. What did people really ***do*** when they made an observation, conjure up a hypothesis, proved a theorem, performed an experiment?

(4) Furthermore, for too long the strenuous debates over the epistemological status of the abovementioned 'contextual' approach divided scholars between those who drew attention to the so-called '*intellectual* factors'- concepts, methods, empirical data – and those who stressed '*social* factors' – forms of organization, sociocultural

and economic influences, etc. However, the gap is taken for granted now as 'a rather silly demarcation' (Shapin 1996). There is as much society 'inside' the theoretician's office, classroom and the scientific laboratory, and internal to the advancement of scientific knowledge, as there is 'outside.' Hence the important task for the sociologically minded historians and philosophers of science is to exhibit knowledge making as *social process*. Therefore one should not be amused when in the course of science advancement "he who has the most, and the most powerful, allies wins" (Shapin & Schaffer 1985, p.409). Moreover, the individuals participating in the production of scientific knowledge are historically, geographically, and socially situated; and their reasoning inevitably reflects their situations. This fact should not undermine the *normative* enterprise of philosophy of science, but requires its expansion, by hook or crook, to include within its scope the social interactions within and especially between scientific communities (Longino 1990, 2002).

(5) Since there is no such thing as 'the essence' of a scientific revolution, a plenty of stories of scientific advancement can legitimally be told, each aiming to depict some peculiar features of the past. Indeed, what one selects inevitably represents her own interests, even if she aims to 'tell it like it really was.' This is especially true when one looks on past from the modern point of view, trying to elicit in the past methods and ways of dealing with the problems that can be useful in treating with modern problems. The most vivid example is modern problem of unification of general relativity and quantum field theory which goads to plunge into Einsteinian, Maxwellian and Newtonian syntheses, etc.

(6) Careful observations and data taken by scientists are not by themselves confident evidence for or against any particular hypothesis. On the contrary, the relevance of any particular data for any given hypothesis is determined by human beliefs and proper assumptions about what kinds of data can support what kinds of hypotheses. Nevertheless, the existence of multifarious perspectives to criticize hypotheses can turn some of the hypotheses into respectable scientific knowledge. Hypotheses become scientific

knowledge when they are subjected to scrutiny from various perspectives, especially by those with *diverse* beliefs and values (Longino 1990).

(7) The Principle of the External Standard can be applied to shift from one practice to another. It considers a shift in practice as rational if

> "the process through which the shift was made has a success ratio at least as high as that of any other process used by human beings" (Kitcher 1993, p.303).

(8) The cognitive processes that eventuate in scientific knowledge are themselves social first and foremost in the following sense (Longino 1990, 2002). The semantic gap between statements describing data and statements expressing hypotheses or theories to be confirmed (or disconfirmed by that data) is created by the difference in descriptive terms used in the description of data and in the expression of hypotheses. The gap means that evidential relations cannot be formally specified and that the data cannot definitely support a theory or an hypothesis to the exclusion of all alternatives. On the contrary, such relations are mediated by certain background assumptions. And eventually, in the chain of justification, one reaches assumptions for which no evidence is available. If these are the context in which evidential relations are constituted, questions arise concerning how the acceptance of such assumptions can be legitimated. According to Helen Longino, the only check against the arbitrary dominance of subjective (metaphysical, political, aesthetic, etc.) preference in such cases is critical interaction among the different groups of the scientific community or better among the members of different communities. Longino considers the underdetermination argument as expressing in logical terms the point made by the sociologically oriented researchers. Viz., the individuals participating in the production of scientific knowledge are historically, geographically, and socially situated and their observations and reasoning fatally *reflect* their situations. This important fact does not undermine the normative enterprise of philosophy of science, but necessarily requires its expansion to include within its scope the social interactions within and

especially between multifarious scientific communities. (See the Planck-Einstein chapter of this book). What counts as knowledge is determined by such interactions also.

(9) *Monism vs. Pluralism Debates. Monism* (Kellert, Longino, and Waters 2006) holds that the goal of inquiry is and *should* be a unified, comprehensive, and rather complete account of phenomena. If this is really so, there should be *one* standard by which theories, models, and even hypotheses in the sciences are assessed. Inevitable deviations from an accepted theoretical framework are problematic and require explanation. It is important that monism, with its commitment to ultimate unity, requires ways to reconcile competing theories, research programs, and paradigms or to adjudicate controversy so as to eliminate competition in favor of the one true or best theory.

Pluralism holds that the observed plurality of approaches within a science is no decisive disadvantage but rather reflects the complexity of the phenomena under investigation in interaction with the limitations of human cognitive capacities and the variety of human cognitive as well as pragmatic interests in representations of those phenomena. For instance, Cartwright's (1999) concept of a 'dappled world' stresses the complexity and diversity of natural and social worlds. Different theories and models are representations of varying degrees of abstraction that manage to apply at best partially to whatever phenomena they purport to represent.

Likewise, Kellert, Longino, and Waters (2006) see no reason to require that the multiple accounts must be compatible. The multiplicity of noncongruent empirically adequate accounts helps us appreciate the complexity of phenomena without being in a peculiar position to generate a single account of that complexity. They insist that *pragmatic* considerations will govern the choice of theory/model to be used in particular circumstances.

Nevertheless, the monist stringent standpoint leading to the reconciliation of the traditions, programs and paradigms encountered seems to me *more preferable* due to the following reasons.

(9.1) The reconciliation strives to *eliminate* the cross-contradiction between the paradigms; yet even Henry Poincarè had demonstrated that the drawback of the theory grounded on inconsistent foundations consists in that *any* consequence follows from it. The holy religious texts are the most vivid examples.

(9.2) It is Fundamental Theoretical Scheme (FTS) of a mature theory structure that can meticulously describe the *relations* between the partial theoretical schemes relating to different subjects (see Stepin 2006 and Chapter 3). For instance, Maxwell's contrivance of electrodynamics famously resulted in getting the $c= 1/\sqrt{\varepsilon\mu}$ equation connecting the speed of light c with the properties of the medium ε and μ.

(9.3) Only the reconciliation model can theoretically reproduce the apparent processes of the *interpenetration* of the methods and the results of the research programs common for the history of science. (Recall the histories of classical mechanics, Maxwellian electrodynamics and special and general theories of relativity depicted in Chapters 1, 3, 4, and 5).

Thus, the author's problem consists in taking into account the abovementioned sociology of knowledge results in conjunction with certain 'internalist' advancements concerning mature theory structure (Stepin 2006; Worrall 2000, 2020; Cheng 2022). Ultimately, the aim of the book is to contrive such a comprehensible *ideal model* that provides analytical tools for studying the mechanisms by which the cognitive structures of scientific knowledge evolve and interact with each other and with social dimensions of the historical advancement of science. And to exhibit how the model 'works' reconstructing the numerous case studies.

In the first approximation, the desired model is reduced to the following abstract provisions.

The cognitive aspect is really characterized by a proposition that science develops due competition of different research programs, traditions, etc. But the studies revealing the social facet insist that a SRP should be taken as a bundle of multifarious theoretical and empirical/experimental practices united by a certain 'metaphysical hard core.' In the subtle case of modern science the latter is frequently exchanged by a Fundamental Theoretical Scheme typified by a system of abstract objects or basic idealizations (material point, charge, etc.). These idealizations govern the directions of deployment of theoretical and empirical traditions of a mathematical projections of nature. Hence, in reconstructing the so-called 'mature theory-change' one should speak not about SRP's, but about 'mathematical research projects' (Heidegger) each consisting of a network of interconnected practices (Rouse).

However, these rather abstract 'first approximationэ of the model still needs to be clarified and refined by referring to specific episodes in the development of science, which alone can guarantee its effectiveness. And the first thing that comes to mind is, of course, the Copernican revolution.

I.8. A Blunt Origin of the Copernican Revolution: in the Wake of Koyrè, Latour, Husserl and Heidegger.

The near-by goal of the present chapter is to strengthen efforts in landing the sky-high Lakatos-Zahar account in the wake of the abovementioned approach. One should take a further step on the thorny path of inquiry for the mundane and empirically-verifiable reasons for the incipience and triumph of the Copernican breakthrough program. Forsooth, in our humble opinion, the discussed above accounts (I) - (V) unfortunately miss a substantial point of the Ptolemy-Copernicus intricate transition. Viz., the alternative ontologies of Ptolemy (mathematical astronomy) and Aristotle (qualitative physics), all appearance, arose out of diverse social *practices*. Ptolemy's wayward ontology was grounded in pagan religious meditative practices (the divine mathematics of the Skies), while Aristotle's solid worldview was entrenched in sensual and object-manipulative ones (the mundane physics of the Earth). For many centuries, the dualism had been 'frozen up' in the Aristotle-Ptolemy cosmology until the Christian canon Nicolaus Copernicus blew the pagan cosmology up from within. The explosion was performed due to quite different (monotheistic) socio-cultural context. As a result, Kepler and Galileo set up a throng of corresponding research practices aimed at the efficient reconciliation of mundane physics of the Earth with sublime mathematics of the Skies. To sum up, *Copernican and Ptolemaic congenial programs were realizing the alternative ways of astronomy and physics stout reconciliation.*

Withal, let us recall what Paul Feyerabend, who devoted many years to the meticulous study of the first scientific revolution, wrote about the lessons of the Ptolemy-Copernicus breakthrough. *Not single* reason *and not one method, but different* reasons, withal assessed from *diverse* positions, *is what had made up the intricate Copernican revolution.* We would willingly add that these sober reasons and vantage points were relentlessly *intertwined.* Regrettably, his interweaving was substantially random, whereupon one should *not* try to explain the *whole* medley only by the blunt influence of simplified perspicuous methodological rules.

Consequently, the ultimate aim of the present account is to propose merely a *more stout* (but not the ultimate!) answer to the stale question "*Why did the Copernican bold program eventually supersede the entrenched Ptolemaic one?*" To propose a more apt and sober explanation one has to provide a substantial 'theoretically progressive problemshift' relative to other rival reconstructions taking the social-cultural facets into account to a greater extent. And to exhibit again that the Copernican Revolution, alas, is a more inextricable phenomenon than seems from the spellbinding epistemic conceptions of scientific revolutions (Kuhn 1977; Lakatos 1978).

In good sooth, the previous alluring yet scant accounts have intensely oscillated between *two alternative extremes.* On the one hand, in the taken for granted blunt vein, the apparent differences between research traditions were hastily taken to be insignificant and fruitful communication (and even deep interpenetration) unproblematic. On the other hand, in the new-fangled, post-Kuhnian distinctive disquisitions, really significant differences between the research traditions are commonly exhibited to be so radical that their actual communication is regarded almost impossible.

The present account humbly stems discreetly from a more common and mundane 'unhagiographic' *intermediate picture.* Respectively, I readily admit that the substantial differences between the research traditions existed at the various levels, ranging from entrenched ontological commitments and inextricable epistemological beliefs similar to Ptolemy's notorious instrumentalism up to methodological principles and subtle mathematical technique.

However, these often antagonistic traditions were able to *communicate* fruitfully in the creative acts of such pundits as Ptolemy, the Moslem solid astronomers, Copernicus, Galileo, Descartes, Kepler, and

Newton. Owing to the *intersubjective* nature of scientific communications, the research practices had vigorously interacted by the various ways that permitted comparisons, adaptations, and even fruitful *cross-fertilizations*.

The intermediate humble approach first and foremost originates from the consecutive criticism of the narrowness of advanced epistemological models: they as a matter of fact (yet due to the different philosophical reasons) lack the subtle mechanisms of the paradigms' (or SRP's, or research traditions/practices) intense *interactions* (Nugayev 1985 a, b). To meet the sound critical arguments, a lucid 'mellow theory-change' model is to be advanced (see the next chapter for details).

Respectively, the profound origins of scientific revolutions are being fathomed now not so much in the stale discrepancies of entrenched mellow theories with the 'hard facts' (common to all SRPs or 'paradigms'), as in the multifarious collisions of 'old' pivotal research practices with each other. The latter transpire first and foremost in relentless contradictions that *can* be most effectively (yet not always!) eliminated or even resolved in a more general ('global') research tradition. Whereupon, the leading roles in mature theory change are played by the intense 'trade' of the proponents of the old paradigms that may lead to mutual accommodation and even profound interpenetration of the participants' s views. 'In this mutual adjustment lies the germ of their further development into a complete unity' (Max Planck).

Withal, impugning Alan Sokal's legal critique, Bruno Latour had stressed that what Sokal failed to grasp is that "[hard] facts remain robust only when they are supported by a common culture, by institutions that can be trusted, by more or less decent public life, by more or less reliable media." Hence with the rise of alternative 'hard facts' whether or not a proposition is believed depends far less on its veracity than on sundry *conditions* of its 'construction.' Viz., *who* is making it, to *whom* it is being addressed, and *from which* social institutions it emerges.

Yet the inevitable encounter of the stout programs, traditions or 'practices,' their deep interpenetration, and strong twisting provide the erection of a vast *hybrid* realm at first with a haphazard array of crossbred theoretical models. Though gradually, on consecutive soothing and eliminating the various contradictions between the hybrid models, the global crossbred solid system is sedulously set up.

Up to a point, the abovementioned lucid model can be taken as proffering a reliable *mechanism* of practically simultaneous drastic transformation of the 'old' paradigms. The hallmark of the sophisticated mechanism consists in the contrivance of the crossbred systems from all the basic theoretical objects of 'old' mellow theories. The crossbred systems constitute cohesive channels through which the accommodation of the encountered 'old' practices inevitably commences. The accommodation gradually leads to the installation of a throng of diverse *crossbred* theoretical schemes. The hotchpotch will probably be ultimately processed and subsequently generalized to set up a *structure* of a novel global theory. Incidentally, a mellow scientific theory may be empirically successful if it constitutes a kind of a neo-Kantian 'draft' that can be furnished and explicated by the results of new bold experiments that increase its empirical content significantly. The global theory may become such an effective vehicle for the production of new testable statements. For this purpose, it should merge and intertwine the 'old' classical research practices in such a subtle way that they once and for all cease to contradict each other. So that the newly-constructed synthesis can successfully explain and anticipate novel experimental evidence.

Hence we contend that *profound breakthroughs in science were first and foremost not due to ingenious contrivances of brave novel paradigms or bold invention of startling new ideas 'ex* nihilo.' (The gist of the Copernican program was known even to Aristarchus). On the contrary, *the profound breakthroughs were caused, among other substantial things, by the harrowing humble processes of piecemeal accommodation, gradual interpenetration, and discreet intertwinement of the 'old' pivotal research practices preceding such radical breaks.*

Whereupon, sagacious identifying and dexterous resolving the relentless paradoxes revealing the profound clashes between the entrenched research practices turn out an indispensable part of the Scientific Method as such. A case of diverse programs' inevitable encounter leads to a wonted situation when a domain of *hybrid* models emerges formed by plain conjunctions of the models of often alternative research programs (Ptolemy, Ticho Brahe). However, commonly the hybrid models transpire to be self-contradictory; and when this is properly realized (by Copernicus, Maxwell, Einstein, Dirac), the crossbreeds are deftly constructed from the basic objects of *all* the cross-theories. *The successful contrivance of new mellow theory commences owing to the crossbred domain's gradual growth.*

The present account strives to expound that the Copernican breakthrough turns out as an artificial result of elucidation and (partial) resolution of the profound dualism, of the 'deep abyss' (Alexandre Koyrè 1957) between Ptolemy's deft mathematical astronomy and Aristotelian descriptive qualitative physics. Whereupon, it is not accidental that the dazzling writings of Copernicus, Galileo, Kepler, Descartes, Newton, and their exalted disciples were all the indispensable levels of mathematics descendance from Divine heavens to sinful Earth and the reverse drastic conquest of the Sky by Earthly mature physics.

> "It became easier to see Copernicus as superior to Ptolemy once astronomy was no longer seen as simply a navigational device (in which case starting with oneself on Earth as the center of calculations makes sense) but as a branch of physics, understood as a universal science of material motion" (Fuller 2020, p.4).

Note that this process had inevitable socio-cultural roots. Rationalization that took place during the first scientific revolution was first and foremost of the sight, when the Renaissance artists had advanced a subtle grammar and syntax for *quantifying* natural phenomena in pictures (Latour, Edgerton, Ivins).

I.9. The Rise and Advancement of Ptolemaic Ingenious Research Program.

On the grounds of SRP methodology Lakatos and Zahar carefully revealed how Ptolemy and Copernicus had invented and coherently advanced the diverse *research programs*. It turns out that both rival designs had branched off from the same Pythagorean – Platonic 'protoprogram.' According to its basic principle, since heavenly bodies are immaculate, all the astronomical appearances should be masterfully 'saved' by the combinations of *as few uniform circular motions as possible.* The stout principle was set up as the firm cornerstone of the powerful *heuristic* of both rivals; wherein the '*heuristic*' was principal, while the '*hard core*' only subordinate.

The ingenious 'hard core' of Ptolemy's program was made up from the geocentric solid hypothesis in conjunction with the Aristotelian natural philosophy, with its notorious paraphernalia of natural and violent motions and the impenetrable and stiff demarcation line between the shabby terrestrial (sublunary) and sublime celestial realms. Note that Aristotelian physics was an empirical science *par excellence* that imbibed common experience much more naturally and fully than the rigid, strict, elevated, yet abstract science of Galileo, Descartes, and especially Newton. Everyone knows that all the hard bodies quite naturally fall down, while fire facilely soars up.

Moreover, inertial motion is not an experimental fact: common experience obviously contradicts it. Moreover, everyone knows for sure that in Nature 'there is no void existing separately' (Aristotle 2007a). The Sun and the Moon rise and set, while the thrown bodies do not uphold their rectilinear motion.

Due to common experience, immaculately fixed by the flexible notions of Aristotelian natural philosophy, the distinctive '*lebenswelt*' (Edmund Husserl) in which one lives and acts mercifully is not a mathematical one nor can it be properly 'mathematized.' This kind of common reality is the wonted realm of uncertainty, unsteadiness, inaccuracy, etc. exhibited by the vague terms 'almost,' 'a kind of,' 'more or less,' etc.

Whereupon the Greek dazzling thought could hardly concede that exactness can successfully survive in the wry world around us. Likewise, the ponderous matter of stale sublunary world with its inextricable hotchpotch of four basic elements (earth, water, air, and fire) can not adequately 'represent mathematical entities' (Koyré 1957) similar to the ideal objects of Euclidean geometry. According to the authority of Stagiritul, 'the [immaculate] mathematical exactness should be demanded only for the [peculiar] objects lacking matter' (Aristotle 2007b).

On the contrary, the heavens consist of an entirely different, odd, and immutable substance, the celebrated 'aether' (or the so-called 'fifth element'). Heavenly bodies are the components of ether spherical shells that fit tightly around each other in a strict order that commences with the Moon, and extends to the sphere of distant fixed stars. A heavenly shell is characterized by its specific rotation, which accounts for the peculiar motion of the heavenly body contained in it. Though outside the utmost sphere of the fixed stars, the almighty *prime mover* is solidly situated. It diligently imports motion from the outside sphere inward. And the 'natural motions' of all the heavenly bodies are unforced, perfectly circular, and neither speeding nor slowing down.

Consequently, the perfect motions of the stars take their place in accordance with strict geometrical laws. Thus, thanks to Stagiritul's epoch-making doctrine, *mathematical astronomy is possible while mathematical physics is not.* Whereupon the Greek astronomers not only masterfully applied immaculate mathematics but with amazing patience and magnificent skill observed the skies. However, they had not even dared to mathematize inextricable terrestrial motions.

The celebrated Claudius Ptolemy (87-150 AD) reached the acme of ancient science not least because he was a *Hellenistic* polymath: an astronomer, astrologer, mathematician, geographer, and even poet rolled into one. His masterpiece – the '*Almagest*' (an Arabic world) - dominated the European thought for more than fourteen centuries. The socio-cultural context of the Ptolemaic research tradition is transpired by the important fact that the author of '*Almagest*' was a successful resident of Alexandria, the splendid capital of Hellenized Egypt. Hellenistic distinctive civilization was a blend of stunning Greek culture with discreet civilizations of Egypt and Babylon and the corresponding spiritual and material practices.

It constituted a substantial departure from the earlier arrogant Greek attitudes towards the dubious 'barbarian' cultures. The extent to which genuinely hybrid Greco-Asian cultures emerged is rather contentious. Yet the encounter of the three pivotal cultures captured even the elevated domain of theoretical astronomy (Neugebauer 1975).

Unlike the 'antediluvian' Babylonians and Egyptians, who had sedulously examined the heavens merely to keep track of their shabby seasons, the 'sophisticated' Greeks had apprehended astronomy from a sublime theoretical standpoint: they strived to comprehend the basic nature and makeup of the whole Universe.

Claudius Ptolemy, as a typical Hellenist, *craved* (not without success) *for balancing himself discreetly between the three pivotal cultures* in different domains of research activity but first and foremost in ontology and epistemology. In natural philosophy, he deftly oscillated between sophisticated mathematical models and blunt, empirically-grounded qualitative physics, while in epistemology – between the blunt realism and forthright instrumentalism.

On the one hand, he was certainly aware yet decidedly skeptical of Aristarchus' heliocentric hypothesis due to rational arguments grounded on the stout and well-grounded empirically principles *of Aristotelian physics.*

On the other hand, Aristotle, all appearance, turned to be the only philosopher to whom Ptolemy referred explicitly. Hence the author of the '*Almagest*' was many times and oft accused of slavish adherence to the blunt principles of Aristotelian natural philosophy.

However, this dictum is an obvious oversimplification. One should recall that, according to Aristotle, 'the rest is nobler than the motion,' and the Sky is much more perfect and nobler than the Earth. Hence, the '*Almagest*' subtle model of the solar system did *substantially* deviate from orthodox Aristotelianism at least in the two substantial respects.

(i) Aristotle persistently maintained that heavenly bodies should move around the Earth in *single* uniform circles. Yet, in the Ptolemaic ingenious models, the odd motion of the planets is an intricate *combination* of *two* circular motions. Moreover, at least one of them is *non-uniform*.

(ii) Aristotle also asserted – alas, again from the elevated and abstract ontological grounds – that the Earth should be located at the *exact* center of the Universe. Yet in the Ptolemaic feasible and deft, distinctive system, the Earth is decidedly *displaced* from the center of the Universe. Furthermore, in Ptolemy's inextricable cosmology the unique center of the Universe is lacking. The central orbit of the Sun and the planetary local deferents all have slightly *different* geometric centers, none of which coincides with the Earth. In the '*Almagest*' and in other substantial writings as well, Ptolemy had frequently demonstrated that the non-orthodox (concerning the Aristotelian natural philosophy) aspects of his subtle models all were directly dictated by stubborn facts of observations.

Ptolemy's most controversial invention was the notorious '*equant*': a planet revolves around the Sun at a significantly non-uniform rate. However, it can facilely be ascertained that the non-uniform rotation of the radius-vector connecting the planet to the Sun implies a uniform rotation of the radius-vector connecting the planet to the 'equant.' The latter is the subtle point directly opposite the Sun relative to the geometric center of the orbit (Fitzpatrick 2010).

It is merely from the modern historically biased standpoint that Ptolemy's equant turns out a precursor of the celebrated Kepler ellipses. Yet for such caustic monotheistic critics as the Islamic /Arabic astronomers and subsequently Copernicus and his associates[2], the compelled insertion of the equant was a typical Lakatosian 'ad hoc$_3$ hypothesis.' It obviously contradicted the 'Gesunder Verstand' of the Aristotle - Ptolemy program – the pivotal tenet of uniformity of motion in respect to the center of the Universe.

However, according to Lakatos & Zahar's sober stance, the 'ad-hocness' should be taken, within the flexible frames of the SRP methodology, not as a narrow property of an isolated hypothesis but as a multilateral *relation* between two consecutive theories. (Even Ibn al-Haytham (965-1040) initiated the first sober critique of Ptolemy's physics for the violation of the basic principle of uniform circular motion).

It is no coincidence that after Eudoxus' primordial model (made up of the system of concentric rotating spheres) was resolutely abandoned, actual progress in the realization of geostatic program ran *counter* to the powerful heuristic of Platonic 'protoprogram' (Zahar 1973). The eccentric resolutely displaced the Earth from the center of the circle. Furthermore,

[2] the works of the Maragha school might be known to the author of '*De Revolutionibus*' – see, for instance, Swerdlow & Neugebauer 1985

the Apollonian and Hipparchan deft epicycles provided that the actual path of the planets about the Earth was *not* circular. Eventually, the Ptolemaic notorious equants entailed that even the motion of the epicycle's empty center was not simultaneously uniform and circular. The insertion of the equant was the strongest blow upon the stale heuristic of the Platonic protoprogram: it was tantamount to its full wreck. Thus, within the Ptolemaic program *mathematical exactness that insisted on the insertion of noncircular orbits and the centers of rotation not coinciding with the earth center began to diverge irreversibly from the stout and empirically sound principles of Aristotelian physics.* Hence one can take *Ptolemaic wayward cosmology as a substantially* ***dual*** *fundamental theoretical scheme mixing the basic principles of 'Platonic immaculate mathematics' with those of 'Aristotelian sober physics.'*

To retain the indubitable advantages of the Aristotelian doctrine and the conspicuous achievements of the mathematical astronomy, Ptolemy in the '*Planetary Hypotheses*' had to advance further and at the same time attenuate the blatant rupture between the mundane and celestial phenomena. Since the '*Almagest*' confined itself to subtle mathematical models, Ptolemy had to contrive preliminary physical models for the same constructions, yet in vain. An apparent example of his numerous failures is represented by the attempt to invent the ingenious 'tambourine-like,' 'sawn-off' mechanism able to transform motion from outer spheres of his cosmos to inner ones (Murschel 1985).

However, *after the 'Planetary Hypotheses' the inmost abyss between the physical and mathematical facets became even deeper* (Jones 2005). Not by chance Ptolemy's inextricable cosmology could not help but be exposed to repeated attacks during the European Middle Ages (Linton 2010). The Aristotle-Ptolemy pagan mélange apparently confronted the strict principles of monotheism not admitting the stiff demarcation line between the celestial and mundane worlds; *all seemingly different worlds nevertheless should have one and the same Creator.*

I.10. The Copernican Farsighted Breakthrough

All appearance, the ultimate motive of the heliocentric program contrivance consisted not at all in sedulous elimination of the wonted discrepancies between the Ptolemaic 'wry' cosmology and stout observational 'hard facts.' According to Alexandre Koyrè,

> "it is rather strange that astronomical revolution, not only as regards its origin – the observational data of Copernicus are almost the same as those of Ptolemy – but also as regards its development, was almost *entirely independent* of the development of observational astronomy" (Koyrè 1973, p.9).

In my humble opinion, Copernicus was invigorated predominantly by aesthetic and metaphysical (essentially *theological*) considerations. Their far-reaching intention was to effectively eliminate the profound rupture between refined mathematical astronomy and Aristotelian blunt qualitative physics so eloquently depicted by A. Koyrè (1957). So, he criticized the supporters of Ptolemy for (1) they are unable to maintain the principle of uniform circular motion of celestial bodies; (2) they give us an incoherent and irrational picture of the Universe.

Surely, it did not prevent him from engaging the Aristotelian philosophy; in particular, in applying standard Aristotelian 'argumentative technique' to posit the greater probability of his assumptions over those of his geocentric opponents.

> "All these arguments make it *more likely* that the earth moves than that it is at rest. This is especially true of the daily rotation, as particularly appropriate to the earth" (Copernicus [1543], 1972, *De Revolutionibus*, book I, ch.8, 17).

Certainly it was his Christian Weltanschauung that made the rupture between physics and astronomy especially bigoted. Namely this monotheistic standpoint moulded the staff provided by the three pivotal sources: (i) the cosmological views of Pythagoreans, (ii) Aristotle's fine cosmology (Goddu 2010), (iii) and (last but not least!) its inexorable critique provided by the 'Buridan school' (embracing John Buridan, Nicole Oresme, and Albert of Saxony).

> "Copernicus developed a critical approach to natural philosophy but one that enabled him to modify Aristotelianism while remaining a participant in a broadly conceived Aristotelean

> tradition [...] He developed his own version of Aristotelianism, very much influenced by Renaissance Platonism, Neoplatonism, Stoicism [...] It was also a version of Aristotelianism that was anti-Averroistic and more flexible in its approach to mathematics" (Goddu 2010, pp. xxii, xxiv).

In that respect Copernicus can be taken as a scholar belonging to a respectable ancient tradition of reconciling Ptolemy's mathematical models with the concentric cosmology of Aristotle (Barker & Vesel 2012). Though now the respectable tradition was strengthened by the favorable socio-cultural context of Christian Weltanschauung enforcement.

It is no coincidence that the celebrated introduction to his opus magnus - '*De revolutionibus orbium coelestium*' (1543) - was dedicated to 'his holiness Pope Paul III.' The respectable canon at Frombork Cathedral, Doctor of Theology, whose maternal uncle (and powerful promoter) was Lucas Watzenrode, the honorable bishop of Warnia, openly admits that he have been inevitably

> "impelled to consider a different [from Ptolemy] system of deducing the motions of the universe's spheres for *no other reason* than the realization that *astronomers do not agree among themselves in their investigations of this subject.*"

Furthermore,

> "those who devised the eccentrics seem thereby in large measure to have solved the problem of the apparent motions with appropriate calculations. But meanwhile, they introduced a good many ideas which *contradict the first principles of uniform motion.* Nor could they elicit or deduce from the eccentrics the *principal consideration*, that is, the structure of the universe and the true symmetry of its parts."

On the contrary,

> "we discover a marvelous symmetry of the universe, and an established *harmonious* linkage between the motion of the spheres

> and their size, such as can be found in *no other* way" (Copernicus, [1543], 1972, pp.I, 22).

It was Copernicus' Christian Weltanschauung that grounded his discreet attitude towards the sober arguments of the pagan Aristotle. Therefore, he is not limited to referring to the authority of Stagiritul to justify the unacceptability of uneven movement. Instead, he proffers a weighty metaphysical argument of *his own*: variable speed can only be caused by variable force. However, God, as the Primordial Cause of all the movements, is always constant. Consequently,

> "I began to be annoyed that the movements of the *world machine, created for our sake by the best and most systematic Artisan of all,* were not understood with greater certainty by the [pagan] philosophers, who otherwise examined so precisely the most insignificant trifles of this world" (Copernicus [1543], 1972; my italics).

Secondly, while for old-fashioned Aristotle gravitation constitutes the tendency of heavier bodies to long to the center of Universe, for Copernicus it is merely the tendency of heavier bodies to get to the centers of *any* spherical masses of matter.

The true origin of the inextricable paradoxes, due to Copernicus's distinctive view, consists in the non-ideal odd movement of the planets. However, according to the Aristotle - Ptolemy stiff doctrine, resolutely elevated to dogma by the Christian faith, the planets necessarily belong to ideal spheres and *should* be engaged in uniform motions along with the perfect circles or along with their artful combinations.

Following not so much Aristotle as his own stout theological reasons, Copernicus was convinced that the supposed perfection of the heavens requires celestial bodies to execute *uniform* circular motion only. Hence he was spurred to decisively stave off first and foremost the dubious equant model.

The resolute rejection of the equant was connected with that Copernicus not only employed a great deal of observational data from Arabic astronomical sources, but also many subtle theoretical advances as well. The latter first and foremost included the famous 'Tusi couple,' used by Copernicus to exchange Ptolemy's clumsy geometrical equant construction. Henceforth he willingly agreed with the blunt criticism of Ptolemy's model by Ibn al-Haytham, Ibn Rushd, Nur as-Din al-Betrugi and other Moslem conspicuous astronomers (Kokowski 2012).

In his magnum opus Copernicus realizes the task targeted first in the '*Commentariolus.*' The initial motivation for contriving the heliocentric system consisted in the assertion, that if the planetary orbs are ordered around a single center according to a strict principle, then the Earth cannot be that center. Invigorated by the '*best intentions to layout the Divine Order of the Heavens*' Copernicus strives to return to Aristarchus's queer proposal to place the center of the Universe on the Sun. But this attempt necessarily generated the paradoxes within the Aristotelian physics inextricably connected with the stale notions of 'natural' and 'violent' movements. As a result, Copernicus had gradually contrived merely a sophisticated *crossbred* theory. It incepted to pave the way to divine immaculate mathematics and mundane qualitative physics deep *interpenetration.* As outstanding French historian punctuated,

> "Copernicus in an insinuating manner and probably unconsciously had inserted into the stout Aristotelian fortress two innocent premises through which Kepler, Galileo, and Descartes vigorously blew it up" (Chaunu 1984, p.430).

Just to punctuate. In startling Introduction to '*De Revolutionibus,*' Copernicus passionately appeals to a highly influential *coterie* of respectable clergy including Pope Paul III (to whom the opus magnus was dedicated), Pope Clement VII (who not only adopted the '*De Revolutionibus*' but even strongly insisted on its publication), Nicholas Schonberg, cardinal of Capua, Lucas Watzenrode, bishop of Warmia, Tiedemann Giese, bishop of Chelmno, et al. Throughout the whole chef-d-euvre its author consistently condemns the author of '*Almagest*' for inherent *paganism.* Copernicus's scathing criticism of Ptolemy's stale system is grounded on the weighty argument, according to which the Pagan's inextricable medley system (deftly embracing tens and hundreds of epicycles, abstruse epicyclets, and quirk equants) is apparently lacking strict, monotheistic Order, pre-established by the Lord, 'the best and *most systematic* Artisan.' Incidentally, sometimes abovementioned instrumentalism of the notorious '*Almagest*' went so far that one and the same movement of the same planet was displayed by two *substantially different* mathematical models. (Recall, for instance, Ptolemy's math models of Mars's odd motion). All appearence, the multifarious components of the Ptolemaic motley system vividly epitomize the various plans of different pagan 'artisans,' of warring Hellenistic gods, densely inhabiting both splendid Olympus and the dismal tombs of the pharaohs as well.

Not by chance,

> "their [the Ptolemy partisans'] experience was just like someone taking from various places hands, feet, a head, and other pieces, very well depicted, it may be, but not for the representation of a single person; since these fragments would not belong to one another at all, a monster rather than a man would be put together from them" (Copernicus [1543], 1972).

Likewise, Copernicus selflessly paved the broad way for Galileo's and Newton's mighty mathematical physics. If the robust demarcation line between divine and mundane worlds is lacking since the Earth is just an ordinary planet of the Solar system, the mathematical subtle notions and elevated principles will apply both to its rotations around its axis and the Sun, as well as to all the ponderable bodies moving along its surface. Furthermore, in subsequent Galileo's works, the graphical Aristotelian 'natural' movements had to be decidedly transformed into abstract and sublime 'inertial' ones. Note that, in the wake of Latour's path-breaking writings, that this rationalization, this 'geometrization' of space was not unique. It took place within 'simultaneous transformation of science, art, theory of vision, organization of craft and economic powers' (Latour 1990, 30).

Likewise, Galileo commenced with descending mathematics from the Skies to the Earth, conjuring up a project of 'Mathematization of Nature' (Husserl). However, he was inspired not so much by '*De Revolutionibus*' as, in the all-pervading spirit of the Renaissance, by Plato's '*Timaeus*' (and the corresponding renowned yet highly controversial discoveries provided by the new-fangled telescope). As a true Copernican, Galileo's hero Salviati famously points out in the '*Dialogues*' in his dispute with the stubborn Aristotelian Simplicio,

> "and [as] to the Earth, we try to ennoble it and make it more *immaculate,* striving to *liken it to celestial bodies* and, in a sense, to place it in heaven, from where your philosophers expelled it."

In epoch-making '*The Assayer*' ([1623], 1957) Galileo maintains that

> "philosophy is written in that great book which ever lies before our eyes — I mean the universe — but we cannot understand it if we do not first learn the language and grasp the symbols, in which it is written. This book is written in the mathematical language, and the symbols are triangles, circles, and other geometrical figures, without whose help it is impossible to comprehend a single word of it" (Galilei [1632], 2001, quoted from Burtt 2003, 75; see also Stillman Drake's distinctive translation, 1957, 237-238).

And to calm down the reader by whom the magnificent Book was written, in his introduction to the '*Dialogue Concerning the Two Chief Systems – Ptolemaic and Copernican*' Galileo famously punctuates that

> "he who looks the higher is the more highly distinguished, and turning over the Great Book of Nature (which is the proper object of philosophy) is the way to elevate one's gaze. And though whatever we read in that book is the creation of the omnipotent Craftsman, and is accordingly excellently proportioned, nevertheless that part is most suitable and most worthy which makes His work and His craftsmanship most evident to our view" (Galilei [1632], 2001). Eventually (for faithful Galileo) immaculate "*mathematics is the language with which God has written the Universe*" (quoted from Lial et al., 1995, p.2).

Alexander Koyré insisted that Galileo's free interpretation of Christian theology was inspired first and foremost by Plato's '*Timaeus*'; especially by the whimsical myth of the creation of the Universe. The central figure of the renowned dialogue – the almighty Demiurge, a divine craftsman – punctiliously constructs the stout mathematical order out of the preexistent chaos to put up the Universe (the 'cosmos'). For that ambitious purpose he punctiliously cuts out small triangles to erect four regular solids; and then he artfully applies them to construct real bodies, plants, and even animals out of them (Plato [386 BC], 2000). Furthermore, in '*Timaeus*' the notion of a divine Craftsman was enriched by the notion of pre-established harmony devised by Him.

Thus, Nature becomes empty and ordered: in the process of its creation God discreetly inserted strict mathematical necessity in it. And mathematical knowledge is not merely true but is substantially *sacred* – and even more than the Holy Bible. Forsooth, while there are plenty of interpretations of the Bible, the mathematical truths are stringently unique and should be resolutely kept out of discussions.

At the sake of mathematization, Galileo had decidedly transformed the subtle methodology of natural science and had elevated mathematization, as well as (real and thought) experimenting up to the highest rank of basic scientific methods (Husserl [1936], 1970). Eventually, it made it possible for the Florentine to arrive at the 'principle of inertia' before Newton.

Yet the very opportunity of implementation of mathematical methods in natural science turned out to be grounded on the elevated procedure of *idealization*, or in Heideggerian terms, on the mathematical projecting of Nature. Thus, the exact science of modernity commenced with taking all natural phenomena as more or less adequate approximations of sublime Platonic 'ideal essences.' The latter lack, contrary to Aristotelian metaphysics, profound existence *within* the natural phenomena. They exist *alongside* them as the 'certain limits of infinitely small sensory becoming'; hence they can be freely contrived by the ingenious human mind. And they are the strict relations between the ideal 'quasi-essences' that are immaculately depicted by the exact Laws of Nature. And the relations between the *real* objects (e.g. rods and clocks / tables and chairs) are fixed merely by the *approximations* to the strict laws. Just as Galileo himself had put it many a time and oft, 'the search for essences, in my judgment, is a vain and hopeless kind of pursuit.'"

Consimilar platonic (to be precise, neo-platonic) motives that found their guileless expression in the budding tenet of '*delightful accordance between the [base] Cosmos and the Holy Trinity*' induced Johann Kepler to the strenuous search for the stout mathematical laws stiffly governing the planet motions. Educated in strict Lutheran faith, Kepler had passionately devoted himself to delving into the sacred '*Book of Nature*.'

> "We astronomers are *priests of the highest God* concerning the book of Nature" (quoted from Barker & Goldstein 2001).

Nevertheless, the startling views of Kepler and Copernicus differed significantly. In Copernicus's audacious theory, the planetary motions were impeccably circular. They demanded no causes and took part due to the inertia of the divine bodies. Hence the Sun was *not* the actual center of force. Only Kepler's indefatigable reflections on the true source of planetary motion revealed the leading role of the Sun and buttressed him in unfolding the subtle mechanics of planetary motion .

On the firm grounds of Trinitarian thoroughgoing doctrine, Kepler decidedly took the Sun as the geometric and dynamical center of the cosmos. The startling coequality of Father, Son, and Spirit implied the continuity of the Center, Periphery, and Space of the Cosmos. The gorgeous Sun epitomized God the Father, the distant Stars respectively referred to God the Son, while the wonted planets humbly incarnated the Holy Ghost. Moreover, Kepler plied to reveal an immaculate universal law that would meticulously picture the motion of both the Earth and the planets. His assiduous quest was invigorated by the abovementioned analogy between the base Cosmos and the Holy Trinity.

Kepler boldly took the second resolute step towards the deep interpenetration of mathematical astronomy and qualitative physics. In doing it he elicited the laws roughly breaking the Aristotelian stiff principle of uniform rotation of divine bodies. His perspicuous laws turned out to be the first scientific laws taking math form. Since Kepler, the Skies relentlessly started ruining the qualitative physics of Aristotle. Masterful reconciliation of the divine and sublunary realms decidedly moved Aristotelian physics aside (Kepler [1609], 2005).

The epistemological import of the Copernican-Galilean scientific revolution (unification of the celestial and terrestrial physics) found its fine expression in the title of Kepler's path-breaking '*Astronomia Nova seu Physica Coelestis*' (The New Astronomy of *Physics of the Heavens*).

Certainly, Kepler was not alone in rejecting Aristotle's natural philosophy from the stringent standpoint of Christian theology. In 1644 Rene Descartes had published '*Principia Philosophiae*' where he proposed a theory of motion *directly* grounded on heliocentric model. Eventually, Descartes' epoch-making writings epitomized the decisive elimination of the Aristotelean stale system. According to modern historians, it was Cartesianism which resolutely turned the handy heliocentric surmise into a dominant paradigm (Chaunu 1984) .

In fine, Galileo and Descartes provided

> "conversion of the human mind from theoria to praxis, from the scientia contemplative to the scientia active et operative, which transforms man from a spectator into an owner and master of nature" (Koyrè 1957, p.vii).

Isaac Newton's farsighted intention was to reveal the stringent laws that firmly dictate the motion of *both* terrestrial and divine bodies. He had pioneered in eliciting, - thanks to powerful heuristic of Copernicus and Galileo, - that the same force attracted all the bodies to the Earth and compelled the Moon to obediently orbit the Earth.

Sir Isaac deftly amended the 'hard core' of the Copernican celebrated program by efficaciously unifying and then generalizing the partial theoretical schemes of Copernicus, Kepler, Hook, Descartes, and Galileo and coming up to the whimsical conjunction of three laws of dynamics altogether with the mysterious gravitation law. Eventually, in startling science of modernity Aristotelian qualitative 'essences' had been ingeniously replaced with immaculate mathematical abstract objects. Lo and behold the beginning of the "*Principia*."

> "Since the ancients (as we are told by Pappus) made great account of the Science of Mechanics in the investigation of natural things; and the moderns, laying aside substantial forms and occult qualities, have endeavoured to subject the phenomena of nature to the laws of mathematics; I have in this treatise cultivated

Mathematics, so far as it regards Philosophy" (Newton [1687], 1846, 1).

In Newton's stiff methodology the decisive dictum 'to subject the phenomena of nature to the laws of mathematics' constitutes the most robust one. A devoted man of science should while sedulously examining the intricate phenomena of nature, force his sense data in such a dry and suitable for experimental purposes way as to proffer them for stringent analytical treatment.

To quote eventually Hermann Weyl's exposition of Husserl's keen 1913 "*Ideen zu einer reinen Phanomenologischen Philosophie*" in thorough "*Raum-Zeit-Materie*":

> "It is the nature of a real thing to be inexhaustible in content; we can get an ever deeper insight into this content by the continual addition of new experiences, partly in apparent contradiction, by bringing them into harmony with one another. In this interpretation, *things of the real world are approximate ideas.* From this arises the empirical character of all our knowledge of reality" (Weyl [1919];1922, p.5).

CHAPTER II.

A Lucid Mature Theory-Change Model.

The delimitation of subject areas and their design within the boundaries of special branches does not separate sciences from each other, but for the first time only creates at their boundaries the interaction through which adjacent areas emerge. The latter have ***their own dynamics****, putting forward new, often solving sets of problems.*

— Martin Heidegger. Wissenschaft und Verständnis

Probably, as a kind of general supposition, it can be contended that those trends in the history of human consciousness turned out to be most fruitful where manifold ways of thinking had ***encountered****. These ways of thinking are deeply grounded in diverse spheres of human culture, in different times, in the plural cultural milieu, or varied religious traditions. When they actually meet with each other, when they correspond to each other so that an* ***intense interaction*** *between them takes place, one hopes that novel and interesting discoveries will follow.*

— Werner Heisenberg. Physik und Philosophie.

Abstract. According to an outline of a scientific revolution model, proposed in the first chapter, profound breakthroughs in science are caused, among other things, by the harrowing piecemeal processes of accommodation, interpenetration, and intertwinement of the 'old' research practices preceding the radical breaks. In this chapter the outline of a scientific revolution is applied to the case of modern mathematized science. In this case the origin of scientific revolution is constituted by the clash of different 'mathematical projects of nature' with each project represented by distinctive bundle of theoretical and empirical practices governed by the basic idealizations model (the Fundamental Theoretical Scheme). In the case of modern physics a mathematical project of nature is typified by a mature physical theory with its systems of theoretical and empirical schemes described by V.S. Stepin.

Contrivance of new mature theory is by no means the result of a sudden individual insight, an uncanny 'Gestaltswitch,' but mundane payoff of *piecemeal* reconciliation of sundry theoretical research practices. Hence a mature scientific theory constitutes a hierarchy of practices consisting of fundamental and partial theoretical schemes, as well as the level of empirical schemes with the directions of advancement governed by common metaphysical 'hard core' of the project.

A comprehensible ideal model that provides analytical tools for studying the mechanisms by which the cognitive structures of scientific knowledge evolve and interact with each other and with social dimensions of the ahistorical advancement of science is posited. According to the model, an origin of a scientific revolution consists not in well-trodden discrepancies between bold theoretical speculations and stubborn 'hard facts,' but in deadly contradictions of mature theories from 'old' rival scientific research programs with each other. The profound contradictions can be effectively eliminated in a more general (global) theory. The hallmarks of global theory and the reductionist and synthetic ways of its construction, as well as the inexorable stages of its thorny contrivance, are scrutinized. It is contended that the unfolding the cross-contradictions constitute the bifurcation points where the influence of socio-cultural factors is the strongest. This suggests a rather conservative though flexible mode of a scientific upheaval. Seen that way, scientific change is not substantially different from other types of socio-cultural changes.

The model is illustrated in a walk chiefly regarding intricate physics of the early XX-th century, the four 'old' mellow programs in the case being Maxwellian electrodynamics, Newtonian mechanics, Boltzmann's statistical mechanics and classical thermodynamics, and to general relativity genesis. Enticing modern examples, referring to electroweak theory contrivance and general relativity and quantum field theory craving reunion, are also involved.

Keywords: mature theory structure, theory change, profound contradictions, socio-cultural factors, Popper, Kuhn, Lakatos, Feyerabend, Stepin, Planck, Einstein, Hawking, Unruh, Zel'dovich.

II.1. A Downright Introduction: What are the True Reasons of Mature Theory Change in Modern Science?

What are the true reasons for mature theory change in modern science? According to the well-trodden standpoint, a theory-change process takes place because of a smashing 'critical experiment.' The experiment unfolds the blatant inadequacy of the 'old' mature theory. The latter manifests itself either as a blunt contradiction to an unmerciful critical experiment or as a notorious trend to adapt to its results using obsequious auxiliary *ad hoc* hypotheses (Karl Popper).

Commonly accepted 'critical experiment' model or conception is grounded, as a matter of fact, on the so-called '*standard hypothetical - deductive interpretation of a scientific theory.*' (The candid term was coined by Frederick Suppe, [1974], 1977). Those who had upheld this alluring standpoint (a part of epoch-making 'Received View') alleged that all the scientific theories are structurally similar to the interpreted calculi of axiomatic theories of mathematics. Whereupon, a scientific theory can be masterfully reconstructed so that all its diverse content may be logically deduced from a few initial premises. At this juncture, the refutation of any single result necessarily leads to the refutation of the initial hypotheses and, consequently, of the whole theory. Single critical experiment is quite sufficient to destroy the entire gorgeous theoretical edifice.

Yet today the majority of philosophers of science (especially after the untimely demise of logical empiricism) maintain that the standard hypothetical-deductive interpretation of the scientific theory is blatantly insufficient. This forthright interpretation cannot adequately grasp the large group of scientific theories, as these theories contradict accepted standard interpretation of canonical axiomatic reconstruction. Furthermore, as a matter of historical fact, experimentation and theory too often do not change in step with each other, as stressed, for instance, by Peter Galison (1987, 1988).

As a result, the so-called '*nonstandard*' interpretation has replaced decidedly the standard one. Wherein the latter has yet to be fully defined, and its numerous partisans differ in their views significantly. The problem situation resembles the pre-standard one fixed by the following Cassirer's evaluation.

> "Now not only does the picture of nature show new features, but the view of what a natural science can and should be and the problems and aims it must itself undergoes more and more radical transformation. In no earlier period do we meet such extensive argument over the very conception of physics, and in none is the debate so acrimonious [...] When Mach or Planck, Boltzmann or Ostwald, Poincarè or Duhem are asked what a physical theory is and what it can accomplish we receive not only different but contradictory answers, and it is clear that we are witnessing more than a change in the purpose and intent of investigation" (Cassirer 1950, p.83-4).

However, almost all the non-standard concepts share the following strong doctrine.

A scientific theory necessarily embraces *at least three groups of hypotheses* that differ from each other in their basic functions to be dwelled on below.

(i) The *mathematical* hypotheses that constitute the stiff mathematical apparatus of the scientific theory.

(ii) The *semantical* hypotheses that shape the deft models of a scientific theory.

(iii) The *fundamental* hypotheses that arrange the basic premises of a scientific theory.

Besides the third group of hypotheses, some additional, yet less sublime and abstract, hypotheses are assumed during the long strip of theory construction. These indispensable additional hypotheses cannot generally be derived from the basic premises of the scientific theory.

When masterly added to the group of initial assumptions, some of the additional hypotheses may furnish a particular theory, which has a direct relationship only to a relatively small amount of phenomena.

Whence the adepts of nonstandard interpretation share the unequivocal point of view that *no* basic theoretical assumptions can be *directly* compared with experimental/observational evidence. This comparison could be procured only via the hypotheses of the *intermediate* level.

Thus, the Standard Hypothetical-Deductive Model was exposed to be unsuitable, since it had imposed too over-strict constraints on the intricate structure of a scientific theory. Yet it possessed at least one indisputable asset - a *simple clarity*. Yet the forced adoption of the nonstandard interpretation unavoidably broaches the following subject. If no strict rules govern the transition from basic theoretical premises to experimental results, how can a mature theory be refuted as a *whole*? And if it cannot be refuted, then *how* - in the real history of science - did one mature theory replace another? Were the scientists engaged in the breakthrough process spurred by rational considerations? Were they eventually right or wrong?

In the history of the philosophy of science, the holistic standpoint is commonly associated with the so-called 'Duhem-Quine' tenet. As Willard V. Quine reliably substantiated,

> "The *totality* of our so-called knowledge or beliefs, from the most casual matters of geography and history to the laws of atomic physics or even of pure mathematics and logic, is a man-made fabric which *impinges on experience only along the edges.* Alternatively, to change the figure, total science is like a field of force whose boundary conditions are experience. A conflict with experience at the periphery occasions readjustments in the interior of the field. Truth values have to be redistributed over some of our statements" (Quine [1951], 2000, p.345; my italics; see also Duhem [1905], 1954, p.187).

Note that the renowned 'Duhem-Quine' tenet finds substantial epistemological support in a rather diverse social-cultural milieu. Viz., according to one of the sophisticated nonstandard accounts (Stepin 2006), a mature scientific theory (or 'the mathematical project of nature') constitutes a set of propositions that describes the relationships between abstract theoretical objects of at least two types – the *basic* objects and the *derivative* ones. The basic model makes up the array of basic initial idealizations (the Fundamental Theoretical Scheme or FTS) with no direct reference to experimental data.

For instance, the FTS of classical (Newtonian) mechanics is represented by the basic ideal model embracing the set of such apparent *idealizations* as 'the Material Point,' 'the Force' and 'the Inertial System of Reference.' The derivative theoretical objects are formed ('constructed') from the basic ones according to subtle rules. The congenial sets of derivative objects constitute *partial* theoretical schemes or PTSs. A mature scientific theory gingerly expands, to span each novel experimental situation, via the transitions from FTS to PTSs.

Each construction of PTS out of the FTS represents an intricate problem that cannot be bluntly reduced to a strict and trenchant algorithm. Thus, it seems impossible, at first sight, to refute the FTS.

Indeed, if the PTS bluntly contradicts reliable experimental results, another one can always replace it. Then what are the true reasons for mature theory change in science? Elicitation of a simple divergence between theoretical predictions and experimental results refutes the partial theories only but not mature ones. When such anomalies are revealed, PTS can always be modified to coordinate the whole mature theory with experience. If *a* single arbitrary anomaly can always be eliminated, one can just as facilely eliminate two, three, and four and so on – one *after* another. Whereupon how many anomalies are necessary to renounce a mature theory?

Apparently, an *infinite* number of them should be involved. While a mature theory can be adapted to incorporate the results of any *finite* number of experiments, it cannot accommodate an infinite number of them. However, what is meant by the proposition that the theory 'contradicts an infinite number of experiments?' - The fact is that theory contradicts *all* the experiments of a certain type, i.e. not only the experiments that have been performed already but also those of the future. However, it then follows that one mature theory contradicts another! Only genuine theoretical law can definitely predict the results of future experiments, which have yet to be carried out. It is a platitude that a theory is conjured up not only to explain the existing experimental data but also to predict the new ones.

Without claiming to be substantiated, even these downright considerations suggest the dictum that a mature theory can be refuted only when it is impelled to explain a 'critical experiment' on a par with *another* different and contradictory mature theory. The inextricable problem situation arises from the encounter of two mature theories.

The latter, in the inexorable process of their expansion, meet the experiment that belongs to the domains of validity of both theories. An apparent example is the renowned experiments related to blackbody radiation that demanded for their theoretical reproduction and explanation the joint application of dainty Maxwell's electrodynamics, daring Boltzmann's statistical mechanics and immaculate classical thermodynamics (see chapter IV for colourful details).

The other examples are represented by the experiments that produced the empirical basis of quantum mechanics relied on classical assumptions of Maxwellian electrodynamics and Newtonian mechanics that contradicted quantum mechanics (Chang 1995). In all these experiments, magnetic deflection was the common method of measuring the kinetic energy (or the magnitude of momentum) of microscopic particles. The magnetic deflection method of momentum and energy measurement is an effective tool in modern particle physics. Viz., it is employed by the specialists when they examine cloud-chamber or bubble-chamber pictures and infer the energies of particles from the curvature of their tracks. Consider an electrically charged particle entering a region of magnetic field. According to classical theory of electromagnetism, the trajectory of the charged particle is bent to go in a circle with the radius proportional to the magnitude of the particle's momentum: $p = r q B$. (Here p is the magnitude of momentum, r is the radius of the trajectory, q – electric charge of the particle, B – the strength of the magnetic field). This expression is derived from the conjunction of Lorentz force law (Maxwellian electrodynamics) and the second law of motion (Newtonian mechanics). And once the momentum is measured, it can be converted to kinetic energy $K = p^2/2m$. It is important that the theoretical foundation of the magnetic deflection method was provided by the amalgamation of thoroughly alien to each other mature theories of Maxwellian electrodynamics and Newtonian mechanics. Nevertheless, magnetic deflection method was applied as a matter of course in studies of cathode rays, radioactivity, photoelectricity, cosmic rays, etc.

Consequently, the account of wonted interaction of a single 'old' theory with experience is insufficient to comprehend (i.e. to 'reproduce epistemologically') the intricate process of mature theory change. Besides, the intense interaction of several 'old' mature theories should be taken into account. If the experimental anomaly occurs because of the inevitable encounter of several incompatible mature theories, it cannot be withdrawn by common deceptive modification methods. It's effective making away demands the proper resolution of the cross-contradiction. Most radically, this can be done by constructing a more general theory, containing the old mature theories as to its partial and considerably modified ones. Just to quote the startling epigraph to this chapter (Planck [1908], 1960, p.4).

The ultimate aim of the present chapter is to elucidate and advance the above heuristic arguments, showing off an 'ideal' model of mature theory change, and to manifest the model's heuristic asset in the history of physics. (Recall the introduction: a theory-change model is 'ideal' in the same way as 'the ideal model of a gas' embracing a throng of material points and representing blunt approximation to the real situation). Further the contrived model is illustrated chiefly concerning the 'Planck-Einstein breakthrough' in the early XX-th century, the three 'old' mellow theories, in this case, being Maxwellian electrodynamics, statistical mechanics, and classical thermodynamics.

Ipso facto, I hope that the rational reconstruction of this particular episode in the history of science is fascinating in itself. It both elicits a common origin of both the 'old quantum theory' (QT) and the special theory of relativity (STR) and explains why they were contrived almost simultaneously. In some significant respects, the reconstruction supersedes the existing pre-eminent accounts. Further to the stout belief that the 'context of justification' cannot be comprehended outside the 'context of discovery,' I maintain that the STR turned out merely a stage in the startling advancement of the so-called 'old quantum theory.' Whereupon, the true reasons why it was accepted cannot be comprehended without taking into account Einstein's writings (and the startling works of his contemporaries) in the quantum realm (chapter IV). The STR and the old quantum theory were contrived within the same program of Boltzmann's statistical mechanics, classical thermodynamics, and

Maxwellian electrodynamics reconciliation. Or, in candor language, quantum and relativistic revolutions were strikingly simultaneous, since they sprung out of a common origin - the stout clash between the three mature theories of the second half of the XIX-th century that constituted the 'body' of classical physics.

The very implementation of indispensable *reductionist* and *synthetic* research programs is brought about by the clash of different mathematical projects of nature, which they are designed to soothe and even take away. Note that this is the bifurcation point at which sociocultural factors come to the surface since the ultimate rational argument on which way of unification *should* be chosen is lacking. Not by chance within these programs empirically - equivalent scientific theories are settled. Roughly speaking the theories from diverse research programs are empirically - equivalent since they are built up out of the same pile of bricks (see Nugayev 1985a for details) yet in according to diverse architectural projects.

For example, Lorentz's dainty electron theory was worked out within the reductionist program, while Einstein's STR - within a sheer synthetic program of mechanics, electrodynamics, and thermodynamics reconciliation.

Being taken independently, Einstein's theory did not explain any new experimental fact. For apparent reasons (the second STR postulate), the predictions of the Lorentz theory were identical to that of the STR, so that it would not be possible in any case to distinguish between the two rival theories on experimental grounds. Hence most of Einstein's contemporaries had vividly discussed the 'Lorentz-Einstein electron model,' brooded on the 'principle of relativity of Lorentz and Einstein,' and so forth. At the time of publication of Lorentz's dainty, second-order theory (1904) the only data available to test these theories were Kaufmann's (taken as dubious) measurements of the masses of slowly moving electrons. However, they were initially regarded by the scientific community as contradicting *both* STR and Lorentz's theory. It took a whole year for Einstein to retort Kaufmann's paper.

In the end, the program promising to effectively take away the cross-contradiction ultimately prevails. Having tentatively compared the heuristic potentials of the reductionist and the synthetic programs, I favour discreetly the latter group since it has the following *objective* advantages.

(i) Firstly, synthetic programs *should* provide a larger enhanced empirically progressive shift of problems solved than the reductionist ones.

(ii) Secondly, only these programs can rationally explain the application of the so-called hybrid or crossbred objects, which commonly spring from the coincident theories.

If one delves into the multifarious structures of two celebrated theories, - quantum theory and general relativity, – he/she can conclude that their global theoretical schemes sprung out of the subtle amalgamation of the crossbred theoretical schemes. This is in no way to demand that each cross-contradiction *should* be made away (in the stale Hegelian way) by blunt choosing a 'dialectical' synthetic program. I do not assert that *all* the cross-contradictions that took their places in the history of science were cut down at the root with the help of synthetic programs. It means only that synthetic programs have some *objective* alluring properties, and that is almost all that can be said on the intense competition of synthetic and reductionist programs in general. At the same time, this is not to disparage the lead of experiments in science. On the contrary, the theory-change model proposed seems to elaborate further on the current standpoint that *both* theorists and experimentalists have significant breaks in their respective traditions, but they are not typically *simultaneous* (Pickering 1985; Galison 1987). The most vivid argument being that a scientific research program constitutes the mathematical projection of nature. And the latter, by Heidegger's definition, consists of theoretical, as well as empirical practices.

One should recall Einstein's solid statement of 'fact' in frank '*Autobiographical Notes*' concerning the origins of the Quantum Revolution:

> "Fundamental crisis set in, the seriousness of which was suddenly recognized due to Max Planck's investigations into heat radiation (1900). The history of this event is all the more remarkable because, at least in its first phase, *it was not in any way influenced*

by any surprising discoveries of an experimental nature" (Einstein 1949a, p. 37; my italics).

In short, theory advancement must have, up to a point, a life of its own. Or, in Heideggerian terms, "theoretical research is not without a praxis of its own" (Heidegger as quoted from Kochan, 2017, p.67).

The sedate advancement of two pivotal cultures within science does not mean that the two cannot many a time and oft speak to each other.

On the other hand, the issue under consideration can be approached rather differently. In my view, one of the philosophical sources of the developed theory-change model consists in '*internal realism*' amended by the so-called '*coherent theory of truth*' (see Nugayev 2020). The gist of internal realism can be disclosed by its comparison with its obsolete '*metaphysical*' ancestor. The common tenet for both consists in the natural assumption that the world is made up of some mind-independent objects. However, while for stout metaphysical realists there exists only *one* true and complete description of 'the way the world really is,' the internalist pliable philosophers contend with the discretion that there is more than one 'true' theory or description of the world (Putnam 2001). 'Truth,' in tentative internalist approach, is an odd kind of *idealized* rational acceptability. It is some sort of *ideal coherence* of our beliefs with each other and with other experiences as those experiences are themselves embraced by our belief system. Hence, truth is not mere correspondence with a mind-independent or discourse-independent state of affairs. There is no sublime 'God's Eye point of view' that one can stubbornly access. There are only diverse, multifarious (and often antagonistic) standpoints of actual persons reflecting various (often hostile) interests, intentions, values, and purposes.

Consequently, internal realism does not decidedly renounce the stale standpoint that there are experiential inputs to knowledge. Knowledge does not constitute a solid bulk of information with no constraints except internal coherence. Yet internal realism does deny that there are inputs, which are not themselves to some extent shaped by our vague concepts, by the intricate vocabulary we seek to report and portray them, or any inputs, which admit of only one single description, independent of all conceptual choices. What makes a statement (or a set of statements) rationally acceptable is its coherence and fit: coherence of 'theoretical' (or less experiential) beliefs with one another and with closer to experiments beliefs.

Eventually, the history of science can be looked through a fluctuating pattern of multifarious - theoretical as well as experimental - research traditions. Of this entire hotchpotch, only those can, as a rule, survive that manage to match and even support each other.

For instance, the world of 'old,' pre-Einsteinian XIX- th century physics was conceptually and socially fragmented. It was split on four pivotal research traditions belonging to Maxwellian electrodynamics, Boltzmann's statistical mechanics, classical thermodynamics, and Newtonian mechanics. Entrenched traditions organized around diverse groups of phenomena engendered slight support for one another. The stout partisans of each distinctive tradition recognized with a gnashing of teeth the existence of the other but doggedly went their wayward ways. But with the advent of quantum theory and relativity, conceptual reconciliation of worldviews was accompanied by a social reconciliation of practices.

All in all, the guileless idea that runs through the whole book as a red thread is rather straightforward. Alas, the renowned simple and clear accounts that had sought to comprehend mature theory change only on the *monistic* grounds (i.e. scrutinizing the dismal advancement of a single theory, a single scientific research program, a single paradigm, etc.) had unfortunately failed. Hence, one is simply *compelled* to cling to a '*pluralistic*' epistemological background. To fathom the true reasons for mature theory change one should scrutinize the intricate interaction, interpenetration, and intertwinement of theories, research programs, paradigms, etc. Wherein a paradigm consists not only of sublime theoretical propositions but of vague values, whimsical metaphysical premises, and implicit research patterns as well. How the interaction of all these queer components of the witch's cauldron should be perceived and punctiliously depicted? – I cannot procure a general and immaculate answer now. Yet I insist that a proper account of the subtle *dialectic* of the 'old' pivotal research traditions for comprehending mellow theory change is indispensable.

II.2. What Inevitably Makes Mature Theory an Established One.

The indispensable preliminary stage of the cross-contradiction occurrence consists in the independent, self-contained development of several mature theories with domains of validity of their own. By definition, a theory is called a mature theory if all its abstract objects are organized into diverse systems belonging to three distinct levels - the level of basic objects, the level of derivative objects, and the level of empirical ones (Stepin 2005). Hence, one can take mature theory as a bundle of theoretical and empirical practices under the common umbrella of a certain metaphysical hard core (or equivalently a Weltanschauung).

To begin with, the fundamental laws of mature theory describe the relationship between its basic abstract objects. The relationship between the derivative objects is highlighted by the consequences of the fundamental laws. For example, 'the electric field at a point **E**,' 'the magnetic field at a point **H**,' and 'the current density **J**' are the basic theoretical objects of Maxwellian electrodynamics. The dainty Maxwell equations immaculately describe the relationship between them.

Newton's laws describe fine relations between the basic objects of Newtonian mechanics. At the same time, 'the absolutely rigid body,' 'the central field,' 'the harmonic oscillator,' etc. represent the derivative objects of Newtonian mechanics. Certain laws of Newtonian mechanics display the relationship between them: that is, by 'the laws of rigid rotation,' 'movement in the central field,' etc.

The basic objects constitute the solid *basis* of a mature theory. This means that, for a derivative object to join the system of theoretical objects, it must first be punctiliously constructed from the system of basic objects according to certain rules. On the contrary, basic theoretical objects are of *independent* construction; that is, they cannot be constructed from each other.

The solid set of mature theory basic objects makes up the basis, i.e. the stable subsystem of theoretical objects. All the basic objects are apparent *idealizations* and cannot reside as real bodies (like tables and chairs). For example, the material point is trenchantly defined as a body free of dimensions. As for the other basic objects of Newtonian mechanics, it is discreetly surmised that the inertial system of reference can be isolated from external influence. Though in reality bodies that can be *completely* isolated do not exist at all (even black holes & multi universes). However, all the basic theoretical objects can be compared with certain fragments of Nature. The 'material points' may be compared with real bodies whose dimensions can be ignored in the problem-solving process (see, for instance, Landau and Lifshitz 1973). The 'force' can be compared with the actual interactions that considerably change the bodies' states of motion.

The derivative subsystems are *subordinated* to the basic one, but they are independent of each other referring to different fragments of the same domain of validity. Each subsystem is characterized by its own set of notions and mathematical equations that constitutes a special part (section) of the mature theory. For instance, Newtonian mechanics consists of the following independent sections: 'the small-oscillations mechanics,' 'the mechanics of rigid body rotations,' 'the mechanics of movement in a central field,' etc. Each of these proverbial sections is characterized by the subsystem of derivative objects of its own. Each subsystem is a reliable model of a particular type of mechanical motion (the small oscillations model, the rigid rotations model, etc.). Particular laws of classical mechanics describe the relations between the components of the subsystem.

In general, subtle relations between a basic subsystem and a derivative one can be highlighted in the following way. A derivative system is contrived from the basis by a process of *reduction*. It means that a mature theory is deployed not by formal logical, mathematical means only but also through fine *gedankenexperiments* with abstract theoretical objects. The reduction is put into effect by punctilious analysis of the nature of the empirically fixed domain of validity.

This domain can be sifted and 'seen through' the lens of an ideal model, formed by correlations of basic objects. According to the peculiarities of each concrete experimental situation, the diverse-varied constraints may be imposed on the basic system. It enables one to transform the basis into a subsystem of derivative objects. Whereupon, the fundamental equations are then applied to the derivative subsystems to become the partial laws. The informal nature of such metamorphoses converts such an inference into a peculiar problem-solving operation. The solutions to such problems are prudently included in theory at its genesis. To masterful theoretician bothered by applying theory, they serve as helpful *patterns* for subsequent puzzle-solving activity. Each problem is scrutinized in accordance with primary *paradigms* (in Thomas S. Kuhn's 1962 sense).

In classical mechanics, the paradigm examples are represented by the well-known 'derivations' from Newton's laws: the small-oscillations law, the movement in a central field law, the rigid body rotations law, etc. In classical electrodynamics, the paradigm examples are sedulous deductions of the laws of Biot & Savart, Coulomb, Ampére, Faraday et al. from Maxwell's equations.

The construction of the derivative objects from the basic ones enables one to compare theoretical knowledge with experience, to explain and predict the results of real experiments. To this end, an *empirical equation* - an intermediate relation – is derived from the partial law. In this equation, the special constructs are carefully intruded. In contrast to abstract objects, the newly born constructs are no longer idealizations, and they can be compared with real bodies now. These constructs are called the '*empirical objects*' and their systems – artful representations of empirical situations - are called the '*empirical schemes*.' Though empirical objects are not equivalent to real bodies. An empirical object represents not a concrete single body with which an experimentalist operates but merely a *class* of such objects. Consequently, an empirical scheme corresponds not to a concrete experimental situation but to a *type* of such situations. For example, the empirical scheme of the Biot & Savare experiment with a magnetic needle and a conductor refers to an experiment with any current in the conductor and any small magnetic needle beside.

Surely, a mature theory becomes 'an *established* one' when the necessary links between *all* the three levels of its organization are properly installed. It makes possible to apply a mature theory as an effective instrument for making definite predictions (see the corresponding part of this chapter for important details). All the bonds between the three levels of an established mature theory should be sufficiently rigid. The indispensable rigidity allows one to connect a prediction referring to the upper level first with all the underlying levels of a mature theory. Hence, it allows one to construct proper experimental device to check the theoretical prediction.

A novel result, obtained in the course of the advancement of the mathematical apparatus, immediately influences all the levels of a mature theory, all the kinds of practices embraced by its Weltanschauung. Hence, a theory can predict, and the predictions can be verified. Certainly, a mature theory obtains the status of the *established* one only when at least some of its predictions are exhibited to be successful.

II.3. Occurrence of a Cross-Contradiction.

The accomplishment of a mature theory T_1 inevitably gives rise to questions about the relation of T_1's basis, viz., $\{\mathbf{B_1}\}$, to the system of basic objects $\{\mathbf{B_2}\}$ of another mature theory T_2. Are the basic theoretical objects from both mature theories constructively independent? Alternatively, is it likely that $\{\mathbf{B_2}\}$ belongs to a subsystem of derivative objects of T_1 (or vice versa)?

Yet the problem is as old as the mountains. Recall that the alternative ontologies of Ptolemy (mathematical astronomy) and Aristotle (qualitative physics), all appearance, arose out of diverse *social practices*. Ptolemy's wayward ontology was grounded in pagan religious meditative practices (the divine mathematics of the Skies), and Aristotle's solid worldview was entrenched in sensual and object-manipulative ones (the mundane physics of the Earth). For centuries, the dazzling dualism had been 'frozen up' in the Aristotle-Ptolemy pagan cosmology until the Christian canon Nicolaus Copernicus blew the cosmology up from within. The startling explosion was performed in a quite different (monotheistic) socio-cultural context, and, as a result, Galileo set up a throng of corresponding research practices aimed at the efficient reconciliation of mundane physics of the Earth with sublime mathematics of the Skies.

In general, it is impossible to answer the above questions without taking into account the following peculiarities of the derivative objects construction.

(i) The rules for working out the derivative objects from the basis are not clearly formulated algorithms. They are vaguely determined by the problem-solving examples or 'paradigms' prudently included in a mature theory at its origin (Ptolemy's '*Almagest*,' Copernicus's '*De Revolutionibus*,' Newton's '*Principia*,' Maxwell's '*Treatise*,' Einstein & Grossmann's '*Entwurf*,' and so on).

(ii) Application of these rules for reducing the basis to a subsystem of derivative objects takes the peculiarities of empirical reality for granted. However, these peculiarities vary considerably from one domain to another.

(iii) When the physical theories are diverse, the construction rules differ substantially from each other being governed by different paradigms.

Points (i) to (iii) exhibit how difficult it is to expose that T_1 is subordinated to T_2 or vice versa. Therefore, in common scientific practice, the simple conjunction of $\{\mathbf{B}_1\}$ and $\{\mathbf{B}_2\}$ is assumed to constitute a new basis.

The true necessity of examining the interrelations between $\{\mathbf{B}_1\}$ and $\{\mathbf{B}_2\}$ emerges when *both* theories are needed to explain certain experimental data. It is soberly assumed that the experimental data can be described by a system of derivative objects constructed from the basic objects of both theories. Henceforth such derivative objects will be called '*crossbred objects*' or simply '*crossbreeds*.'

Recall how Max Planck famously outlined the important lead of experiments for the construction of the crossbred object in a paper with the meaningful title '*The Relation between Physical Theories*':

> "In physics as a practical science, it has frequently happened and still does happen that two theories, developed independently of one another, come into conflict when extended and must be *mutually modified* to remain compatible. In this mutual adjustment lies the germ of their further development into a complete unity, since the chief purpose of each science is, and always will be, the unifying of all its great theories into one that

will embrace all the problems of that science and afford a solution to all of them" (Planck [1908], 1960, p.83; my italics).

The crossbred objects are constructed from the basic objects of T_1. Hence the crossbred system will be a subsystem of T_1 derivative objects. On the other hand, the crossbred objects are put up from the T_2 basis also. Herewith, the crossbred system will simultaneously belong to the crossbred systems of T_2. Consequently, the partial laws of both T_1 and T_1 will exhibit the relations between the crossbred objects. Incidentally, several 'domains of reality' can exist, the description of which may necessitate the joint application of two mature theories. (For the sake of clarity I'll merely dwell on two theories; in reality, three or even more theories are commonly applied, but all these cases without much labour can be reduced to two-theory ones). Hence, several crossbred systems may subsist simultaneously. The joint application of T_1 and T_2 for solving an important problem will be called, applying in a sui generis way the Podgoretzky & Smorodinsky felicitous notion (1980), '*theories' cross*' while T_1 and T_2 will be referred as '*cross theories*.' The set of statements describing the relations between the crossbreeds will be called a '*crossbred theory*.'

For instance, the successful accomplishment of new-fangled Maxwellian electrodynamics (especially after Hertz's experiments) gave rise to conundrums on the relationship between its basis and the system of basic objects of entrenched research tradition of Newtonian mechanics. The task of theoretically describing a black-body radiation spectrum, the electromagnetic radiation processes, etc. demanded the joint application of both theories. (See Poincare's farsighted lecture course "*Electricite et Optique*," Paris, 1890 and Max Planck's fine 1906 book, p.616).

Let me peer at the conspicuous examples mentioned above more thoroughly.

(a) For the theoretical reproduction of a blackbody radiation spectrum, James Jeans (1905) had dwelled on the system of standing electromagnetic waves in a closed cavity. By taking these waves as a system of harmonic oscillators (viz., by constructing the crossbred theoretical objects), he could apply a well-known law of statistical mechanics (i.e. 'the equipartition theorem'). In this bolt upright way, the dependence of blackbody radiation energy on temperature and frequency was elicited. The crossbred theoretical objects system, the correlations of which constitute a model of black-body radiation, makes up a subsystem of classical electrodynamics (it represents the system of standing electromagnetic waves). On the other hand, the same model constitutes a subsystem of derivative objects of classical mechanics (an odd mechanical system with an infinite number of degrees of freedom).

(b) Lorentz's eminent *Theory of Electrons*, which immaculately explained and predicted a vast number of phenomena referring to the 'electrodynamics of moving bodies,' provides a fascinating example of the genuine crossbred theory. Initially, following the traditions of Maxwell and his partisans (Lenard, Hertz), it was assumed that electric charges could be taken as a kind of perturbations of the ether. This enticing assumption was grounded on the basic idea of Maxwell's electromagnetic theory that displacement current is identical to conduction current. Whence Lorentz was able to successfully represent the density of electric current in the form of electromagnetic field flow through a cover. Nevertheless, owing to the whiff of atomistic ideas, Lorentzian electrodynamics contained the notion of currents but taken as a system of electrons interacting with the electromagnetic field. The system of electrons, as a system of charged particles, is a subsystem of classical mechanics. However, as a throng of electromagnetic field sources, it is simultaneously a subsystem of Maxwellian electrodynamics too.

(c) The queer hypothesis of the atomic structure that assumes the existence of the atomic nucleus was proffered by Nagaoka already in 1905 - before Lord Rutherford's renowned experiments with alpha particles. When inventing 'the planetary model of an atom,' the positive charge was defined as an 'atomic nucleus,' and electrons were taken as being 'on a stable orbit around the nucleus.' The system of crossbred theoretical objects, the correlations of which mould the planetary model of an atom, is a subsystem of classical mechanics, and it is a vital component of the model of 'motion in the central field.' On the other hand, this system constitutes a subsystem of classical electrodynamics (the so-called 'Hertz's oscillator').

The statements from T_1 and T_2 describe relations between the crossbred objects. These objects belong to the subsystems of *both* mature theories. Hence, the operation of crossbred construction is identical to that of endowing derivative objects, belonging to both theories, with *novel properties*. These additional internals of the objects of the first mature theory correspond to a brand new set of relationships transported from the second one.

Systems of derivative objects of a mature theory T_1 were constructed before its encounter with another mature one. A system of theoretical objects is a peculiar summary of corresponding experimental studies carried out *independently* of investigations connected with another mature theory T_2 . Therefore, it is not surprising that, thanks to the construction of crossbred objects, theoretical objects with mutually *incompatible* properties should appear in the subsystem of derivative objects of one of the crossbred theories. In the aforementioned case studies, the appearance of objects with mutually incompatible properties was marked by physicists as a dreadful 'ultraviolet catastrophe' (Paul Ehrenfest's apt notion), the exasperating 'paradox of unstable atom' (Wilhelm Wien's notion), and so on. Let us dwell on those ingenious paradoxes.

(a) As a result of constructing the crossbred system, composing the inextricable model of blackbody radiation, the free electromagnetic field transpired to possess two diverse properties: (1) 'to be a system of standing electromagnetic waves' and (2) 'to be a mechanical system with infinite degrees of freedom.' Indeed, from the mechanical standpoint, the electromagnetic field is a field of electric and magnetic forces that are continuously distributed in space. Such a peculiar system has infinite degrees of freedom in contrast to the finite number of degrees of freedom of a common weighty body. Contrary to the field, the material point has 3 degrees of freedom. Its position is determined by three coordinates while the electromagnetic field at a point is determined by all the electromagnetic forces throughout space, i.e. by an infinite number of magnitudes. Independently of Rayleigh and Jeans, and masterfully applying classical statistics, Einstein revealed that, at an arbitrary but finite temperature, the density of the electromagnetic field energy should be infinite. It is quite natural since at an arbitrary finite temperature the same amount of energy (proportional to temperature) falls on each degree of freedom. However, the infinity of electromagnetic field density is apparently incompatible with the second principle of electrodynamics grounded on statistical mechanics. One can permanently extract energy from the cavity containing such radiation and set the 'perpetuum mobile' of the second kind to motion. Therefore, the property of thermal radiation (1) 'to be a mechanical system with an infinite number of degrees of freedom' is incompatible with its property (2) 'to be a system of standing electromagnetic waves.'

(b) As was later underscored by Albert Einstein,

> "This is the basis [i.e. Maxwell's equations for electromagnetic field plus Newton's laws of motion for material points] on which H.A. Lorentz obtained his synthesis of Newtonian mechanics and Maxwellian field theory. The *weakness* of this theory lied in the fact that it tried to determine the phenomena by a *combination* of partial differential equations (Maxwell's field equations for empty space) and total differential equations (equations of motion of material points), which procedure was *unnatural*. The inadequacy of this point of view manifested itself in the necessity of assuming finite dimensions for the particles in order to prevent the electromagnetic field existing at their surface from becoming infinitely large. The theory failed, moreover, to give any explanation concerning the tremendous forces, which hold the electric charges on the individual particles. H.A. Lorentz accepted this weakness of his theory, which was well known to him, in order to explain the phenomena correctly at least in general outline" (Einstein [1936], 1960, p.306; my italics).

(c) The famous planetary model of an atom was constructed as a system of crossbred objects. As a result, the electron was found to have a new hallmark imported from classical mechanics: (1) 'to move in a circular orbit around the nucleus.' However, like any accelerated charged particle, the electron must (2) *continuously* radiate electromagnetic energy. The energy loss must unavoidably lead to the collapsing of the electron on the nucleus. Therefore, the property of electron (1) 'to move in a stable orbit around the nucleus' is incompatible with the property (2) 'to be a negative charge inside the atom.' The conspicuous paradox, which had been first manifested in the Nagaoka blunt model, was retained in the Rutherford model. Because of the paradox, the arrogant scientific community was rather sceptical about the Nagaoka Japanese dainty model but was forced to accept, with teeth-gritting, the phenomenological model of lord Rutherford because of relentless experimental evidence.

A mature theory necessarily contains expressions regarding the relations between abstract theoretical objects. Hence the objects with mutually exclusive properties transpired in the system of derivative objects, should give rise to mutually contradictory statements in the cross-theories. Bearing in mind Michael Podgoretzky and Jacob Smorodinsky's apt notion (1980), I would like to denote the appearance of incompatible statements, when the theories cross, by '*cross-contradiction.*' The vivid examples of such cross-contradictions of the physics of the XX-th century beginning refer to the blackbody radiation theory, electrodynamics of moving bodies, and to planetary atom theory.

(a) 'Heat equilibrium of radiation with matter exists' (a theorem following from the second law of thermodynamics; see Planck's 1906 lectures on the theory of heat) and 'heat equilibrium of radiation with matter does not exist' (a consequence of Rayleigh-Jeans law, see Lorentz's 1909 celebrated book on the theory of electrons). A young participant at the first Solvay conference (Niels Bohr 1911) recalled,

> "This argumentation, analogous to Jeans' analysis of thermal equilibrium, led, however, to the well-known *paradoxical* result according to which no thermal equilibrium is possible" (N. Bohr, quoted from Heilbron & Kuhn 1969).

(b) "What causes all these difficulties? Lorentz's theory contradicts the purely mechanical notions to which the physicists hoped to reduce all the phenomena of the Universe. Indeed, while there is no absolute motion of bodies in mechanics, and there exists a relative one only, in Lorentz's theory there is a peculiar state corresponding physically to an absolute rest state: the state when a body is immobile relative to the ether" (Einstein 1910).

(c) "The atom is stable" and "the atom is unstable."

A short note.

Some of the abovementioned results (though in a less analytical fashion) were independently obtained by German historian of science Dr. Jurgen Renn. In an important paper "*The Relativity Revolution from the Perspective of Historical Epistemology*" (2004) the author contends that

> "The internal structure of classical physics was largely determined by the emergence of three subdomains with relatively autonomous conceptual foundations: mechanics, thermodynamics, and electrodynamics. At the *frontier of these domains*, a class of problems emerged that involved the conceptual foundations of *more than one* subdomain: the '*borderline problems*' of classical physics, which *generated contradictions* between these different frameworks. The electrodynamics of moving bodies stands for the class of borderline problems in which issues of electrodynamic field theory, such as light propagation, were combined with the mechanical question of the role of moving reference systems. *This combination gave rise to a clash* between the ether model underlying classical electrodynamics and the relativity principle of classical mechanics. *Borderline problems such as these* were at the center of Einstein's 1905 papers and catalyzed the transition from classical to modern physics. They gave rise to special relativity, the light quantum hypothesis, and modern statistical physics respectively. Evidently, their existence was not a matter of Einstein's genius or of favorable local circumstances but the consequence of the long-term development of physics, which also created the conditions of the successful accomplishment of the second phase of the relativity revolution. [...] In the case of the creation of special relativity, Einstein's success depended on his

combining the heritage of mechanics, embodied in the relativity principle, with the heritage of electrodynamics, embodied in the principle of the constancy of the speed of light" (Renn 2004, p.644, my italics).

II.4. A Promising Way of Cross-Contradiction Elimination.

The cross-contradiction results from the construction of the crossbred-objects. To eliminate the cross-contradiction in the most effective and reliable way, the theory T_3 must be contrived. It should deftly comprise both mature theories so that to exclude the very possibility of constructing crossbreeds from the basises of both theories. Theory T_3 will be called a '*global theory*.' According to the proposed theory-change model, there are two potent ways of creating a global theory that bar each other: *reductionist* (R) and *synthetic* (S).

(R) Application of a *reductionist* method of creating a global theory is grounded on the natural assumption that the basises of both cross-theories refer to *different levels* of theoretical object organization. Therefore D_1, the domain of validity of T_1, makes up a part of D_2, the domain of validity of T_2. The basis of T_1 is superciliously called a '*true*' basis. T_1 itself is impudently alleged as a '*fundamental*' theory, while T_2 is humbly christened a 'phenomenological' one. The authenticity of a phenomenological theory should be established by sedulous constructing its basic objects from the sublime basis of a fundamental theory, as well as by proving that its basic laws obediently follow from those of a fundamental theory.

Lastly, the basis of a phenomenological theory obsequiously should take its sordid place as a derivative system of the renowned fundamental theory. The possibility of constructing a phenomenological basis from the basis of a fundamental theory must also be substantiated. The relation between fundamental theory and a phenomenological one should be identical to that between the classical mechanics and one of its parts - the rigid body rotation theory, for instance. The intricate problems of constructing the phenomenological basis from the basis of a fundamental theory are of particular importance for a reductionist program. For obvious reasons, they are called '*the fundamental problems*' (Tisza 1963).

(S) The *synthetic* method of creating a global theory is grounded on not less strong assumption. The basic objects of both mature cross-theories are *constructively independent* of each other, though their basises belong to the same object-organization level. Hence, the exasperating cross-contradiction must be eliminated by creating a brand novel system of *global* objects from which the basises of both cross-theories can be successfully constructed.

The fundamental laws of both mature cross-theories should be manifestly deduced from those of the global theory. Eventually the subordinate basises of T_1 and T_2 should take up their ultimate positions as derivative subsystems of the global theory.

What are the substantial differences between reductionist and synthetic ways of contriving a global theory?

First, the proper realization of the reductionist program rules out the application of crossbred objects. They are by the book, bell, and candle rejected by such a stout program. In contrast, the synthetic program resolutely rules out using the crossbreeds as derivative objects of the cross-theories only. This program takes the crossbreeds as belonging to the *highest* level of theoretical systems organization. Moreover, the implementation of the synthetic program results in the creation of a brand novel system of abstract theoretical objects.

On the contrary, in the global reductionist theory the rules of reducing the basis to derivative subsystems are rather strictly determined by puzzle-solving examples contained in the fundamental theory. In contrast, there are no stiff rules for constructing the basic objects of a synthetic global theory. Therefore a synthetic global theory can be put up by amalgamating consecutive cross-basises according to a new *Weltanschauung* (vivid examples are the contrivance of heliocentrism or the construction of Maxwellian electrodynamics – see Nugayev 2018).

Secondly, synthetic global theory can be worked out using the method of mathematical hypothesis; the thorny path of the electroweak theory being a suitable example (Nugayev 2020). Viz., Yang and Mills (1954) had arrived at the farsighted idea that any field theory (and even the theory of strong interactions) can be modelled following quantum electrodynamics.

In the trenchant language of group theory, gauge transformations that leave the Lagrangian of QED (quantum electrodynamics) invariant belong to U (1) unitary group. Yang and Mills strived to contrive a similar theory that would be invariant under local transformations of the group of strongly interacting isospin symmetries - a special unitary group SU (2). Members related to the isospin multiplets of hadrons performed the lead of the first two members of the Lagrangian, which in QED related to electrons and protons. Now the wave function incarnated a two-component field (p, n), representing a doublet of nuclear isospin. As in QED, this part of the Lagrangian turned out to be non-invariant concerning gauge transformations.

Though if one introduces spin -1 W-particles that form a triplet (W^-,W^0, W^+), then she can introduce such parts describing interactions of W-bosons with nucleons in the Lagrangian that its invariance regarding gauge transformations will be provided.

As a result, a jolly odd theory advanced by Yang and Mills have started to be taken as a model (a 'paradigm') of a whole class of gauge theories, embracing even the theory of gravity (see the corresponding chapter for details). Thus, in the electroweak case, the contrivance of a theory commences with a quest for fundamental equations. Only after they had been ascertained can philosophical interpretation and empirical justification start. For this reason, some of XX-th century ways of creating a theory may differ radically from those of the past.

Implementation of the reliable and resolute ways of cross-contradiction elimination is grounded on two *equally reasonable*, but mutually *incompatible* presumptions. Therefore, these methods must be engaged in three *alternative ideal types* of global theory construction programs - two reductionist and one synthetic. Since several derivative crossbred systems can be contrived simultaneously, several synthetic subprograms can cohabit at the same time. (For instance, Bohr and Einstein's 'old quantum theory' sub-programs were unspooled relatively independently, but subsequently merged into a single and powerful one). Each program creates its own sequence of scientific theories on the solid basis of one of the above natural assumptions. Each theory T_{n+1} from the sequence is contrived by the addition of an auxiliary hypothesis to the preceding T_n. Each aforesaid proposition leads to the assumptions exclusive to the given program. Therefore, following Imre Lakatos (1970, p.119), I will call these basic assertions the 'hard cores' of multifarious reductionist and synthetic research programs. The following statement makes up the hard core of synthetic SRP (scientific research program): 'the basic objects of all the cross-theories are constructively independent.' The opposite is true of the reductionist SRP. Besides, the reductionist 'hard core' indicates which system of basic objects constitute the 'true' basis.

Neither a single crucial experiment nor a sequence of such potent experiments can certainly disclose which SRP kind - reductionist or synthetic - can eliminate the exasperating cross-contradiction efficaciously. Hence, each 'hard core' is irrefutable.

Note that it is this bifurcation point where the influence of external (socio-cultural) factors turns out to be decisive. 'Images of Knowledge' (Elkana 1994), i.e. protracted reflections on what knowledge is or should be like, usually play a decisive role in mediating between ideas and interests present in society, on the one hand, and the stiff body of scientific knowledge, on the other.

This is especially true for Newton's, Maxwell's and Einstein's synthetic programs.

To begin with, one cannot flinch from asserting a blunt theological, monotheistic influence in Newton's textbook case of amalgamation of physics and astronomy. Before Newton, Aristotelian – Ptolemaic pagan cosmology defiantly confronted for a long time the stiff principles of monotheism not admitting the blunt and impervious demarcation line between the celestial and mundane manifold realms.

Yet all the diverse worlds should have the same Creator, 'an impressive Pantocrator in the dome' (Newton 1687). Hence "God said 'Let the Newton be' and all was light" (Alexander Pope).

In Maxwell's more intricate case, the solid influence of Kantian epistemological doctrine urging the inexorable ordering of motley array of electrical and magnetic toy models turned out to be crucial. Maxwell's himself confessed that

> "I am working away at electricity again, and have been working *my way* into the views of heavy German writers. It takes a long time *to reduce to order all* the notions one gets from these men, but I hope to see *my way* through the subject and arrive at something intelligible in the way of a theory" (Maxwell's letter to his father quoted from Campbell & Garnett, 1882, p.105).

This self-explanatory confession should be taken altogether with no less renowned one:

> "in the meantime, I have my usual superfluity of plans… Metaphysics – *Kant's Kritik of Pure reason in German, read with a determination to make it agree with Sir W. Hamilton…*" (Campbell & Garnett, 1882, p.77).

Eventually, Maxwellian electrodynamics was gravely created as a result of the old pre-Maxwellian programs masterful reconciliation: the sophisticated electrodynamics of Ampère-Weber, the subtle wave theory of Young-Fresnel, and Faraday's queer program. Eventually, Maxwell's research program pressed out the Ampère-Weber one because it did organically embrace the conspicuous ideas of this mathematically immaculate program, as well as the valuable presuppositions of the research programs of Young-Fresnel and Faraday. Maxwell's landslide victory became possible because Maxwell's unification strategy was grounded first and foremost on Kantian epistemology (Nugayev 2015).

Viz., Maxwell did put forward as a basic synthetic principle the concept that radically differed from that of rival standpoints by its open, flexible, and *contra-ontological* character. Action at a distance, 'incompressible fluid,' 'molecular vortices' and other contraptions were merely 'contrived analogies' (Hon & Goldstein 2012) for Maxwell, capable only to direct the researcher at the 'right' mathematical relations.

In Einstein's more inextricable case the stiff influence of stout Kantian epistemology was largely mediated by new-fangled 'revolutionary proclamations' of Mach and Duhem. Nevertheless, to comprehend in what subtle ways Einstein's medley 1905 breakthrough papers hang together one has to pay special tribute to his *strive for unity*. Einstein's quirk light quanta and terse STR turn out to be only pencil sketches over a gorgeous oil painting, mere stages of unwinding of Maxwellian electrodynamics and statistical thermodynamics reconcilement program. The stale conception of luminiferous 'aether' was a sheer stumbling block for Einstein's statistical thermodynamics in which the predominant role was played by the whimsical light quanta paper. Hence Einstein was fully aware that his bizarre light quanta hypothesis was too audacious to be taken literally.

So he posited its consequence- the well-known 'electrodynamics of moving bodies' - in a markedly Machian, phenomenological way, and the basic relativity postulate was proposed as organically originating from 'instinctive knowledge.' Though Einstein was influenced by David Hume and Ernst Mach in his criticism of strongly reductionist research traditions of classical physics, when related to creative momenta, his 1905 reconcilement *modus operandi* was initiated by Mach's pre-eminent principle of economy of thought. The latter was taken in the context of the 'Stevin-Mach' doctrine amalgamated with valuable inclinations of the regulative spirit of Kantian epistemology (Nugayev, 2019).

Thereafter, being an Eidgenössiche Technische Hochshule (ETH) student in Zurich, young Einstein had a lucky opportunity to continue his Studium von Kant, Kant's superlative Nachlaß in the winter semester of 1897 at the lecture course on the '*Theory of Scientific Thought*' of Dr. August Stadler, a distinguished neo-Kantian of Marburg school (Einstein 1987, pp. 45-50).

Later the degree of his acquaintance with the works of Kant increased only. For example, in 1918 Einstein confidentially reported to Max Born:

> "I am reading Kant's *Prolegomena* here, among other things, and am beginning to comprehend the enormous suggestive power that emanated from the fellow and still does" (quoted from Born 1971, pp. 25-26).

Furthermore, in his 1936 pervasive "*Physik und Realität*" he contended that

> "One may say 'the eternal mystery of the world is its comprehensibility.' It is one of the greatest realizations of Immanuel Kant that the postulation of a real external world would

be senseless without this comprehensibility" (quoted from Einstein 1954, p. 292).

Likewise, in 1949, eminently reflecting on the basic principles of reasoning in theoretical physics, Einstein avowed that

> "the theoretical attitude here advocated is distinct from that of Kant *only* by the fact that we do not conceive of the categories as unalterable…They appear to be a priori only insofar as thinking without the positing of the categories and concepts, in general, would be as impossible as breathing in the vacuum" (Einstein 1949b, p. 674; my italics).

On the other hand, a reductionist program is stigmatized as 'unable to eliminate the contradiction' only after the convincing demonstration of its incapacity to resolve the fundamental problems. The vivid examples of genuine reductionist programs in the history of physics are indissolubly connected with James Maxwell and Hendrik Lorentz. Viz.,

> "To be sure, Maxwell still tried to interpret his field theory mechanically using mechanical ether models. But these attempts receded gradually to the background following the representation of the theory – purges of any unnecessary trimmings – by Heinrich Hertz so that in this theory the field finally took the fundamental position which had occupied in Newton's mechanics by the material point" (Einstein [1936], 1960, p.305).

As for the Lorentz reductionist program, which was inverse to Maxwell's, final turned out the following Einstein's dictum:

> "Furthermore, there was one consideration which pointed beyond the frame of Lorentz's theory. In the environment of an electrically charged body, there is a magnetic field, which furnishes an (apparent) contribution to its inertia. Should it not be possible to explain the *total* inertia of the particles electromagnetically? This problem could be worked out satisfactorily only if the particles could be interpreted as regular solutions of the electromagnetic partial differential equations. The

Maxwell equations in their original form do not, however, allow such a description of particles, because their corresponding solutions contain a singularity. Theoretical physicists have tried for a long time, therefore, to reach the goal by a modification of Maxwell's equations. These attempts have, however, not been crowned with success" (Einstein [1936], 1960, p.306).

Each subsequent, (n+1)-th version of a reductionist or synthetic sequence of theories represents a more adequate realization of a program than the anterior, n-th version. Each of these sequences verges on a certain limit or *ideal* of the global theory. It is this ideal that governs the direction of development of each SRP ideal type

II.5. Can Any Anomaly Set Forth an Intricate Problem Situation?

Can *any* anomaly set forth a real problem situation? – Obviously not. Otherwise, science could not proceed further rash explanations of one, two or three stale phenomena. What kind of anomalies significantly contribute to problem situations? Or, in our vocabulary, what anomalies testify about the blatant disagreement of several mature theories with each other? - To provide a sober answer it is necessary to inquire into the link necessarily connecting the occurrence of the anomalies with the existence of the cross-contradictions. By hook or crook one should unfold the subtle ways by which the cross-contradiction influences the disagreement of theoretical statements with experimental data.

Recall that, to explain the empirical data, a Partial Theoretical Scheme (i.e. a system of derivative abstract objects) is transformed into an *Empirical Scheme* (ES). The ES objects or constructs differ substantially from the PTS constructs (or the constructs of the Partial Theoretical Scheme) since they are surely *not* idealizations now. They can be directly compared with real objects taking part in actual experiments. (In what follows in this part, I shall use and advance the ideas of the relevant chapter of Vyacheslav Stepin's congenial 2005 book).

For instance, let us dwell on the common situation in Maxwellian electrodynamics with a magnetic needle which is near the conductor carrying a current. To check the validity of the Biot-Savare law, one has to compute the needle's angle of deflection. Though at first sight the Biot-Savare formula has nothing to do with real objects. It discloses the correlations of theoretical abstract objects only. The theoretical objects engaged are an 'infinitely small current' and a 'magnetic field generated by this current.' Therefore the Biot-Savare equation cannot be applied directly to explain empirical data. One should interpret the corresponding quantities as relating to the *concrete* experimental situation first.

That's why a certain consequence of the Biot-Savare law - the empirical formulae - is sedulously derived, and the novel quantities are tentatively inserted into it. They describe the needle's deflection angle and configuration of the integral conductor. The empirical data should be compared with this formula only, but not with the genuine Biot - Savare law. Note that though the components of empirical schemes can be directly compared with real objects, they are not *identical* to them. For instance, a real magnetic needle and real conductor carrying a current possess a lot of properties embracing colour, taste, odour, etc. Yet, while constructing the Empirical Scheme, in the course of empirical practice they are represented merely by the following ones: 'to be guided by the magnetic field,' 'to conduct the electric current,' and 'to have a definite configuration,' respectively. Any ES component can be compared not with a single real object or real experimental device, but with a *class* of such objects or experimental devices. For instance, the 'needle & conductor' clumsy empirical scheme refers to *any* experiment with a conductor and a magnetic needle.

An ES can be obtained from above, as well as from below, i.e. not only as a consequence of a theoretical law but from statistical 'generalization' of empirical data as well. All the components interacting in real experiments - the object of the investigation itself and the corresponding experimental device too - are shown off as having statistically invariant properties only. It means that they are laid out as possessing properties that manifest themselves merely through many times and oft-produced observations. An ES displays the features typical for a lot of experimental situations.

Correspondingly, a subsystem of derivative theoretical objects can be compared merely with an ES class. Therefore a system of derivative objects constitutes an *invariant content* of the empirical schemes. The distinctive notion 'invariant content' should not be comprehended as referring to the notorious *inductive* way of derivation of theoretical laws. To find a genuine theoretical invariant, one should know *beforehand* that the set of empirical schemes certainly makes up a class. This class is easily made out only from above, only via discreet reduction of derivative subsystems to empirical schemes. Yet comparing multifarious empirical schemes with each other, one cannot resolutely conclude that they have common content. The latter is determined by the basis and the rules of derivative objects construction from it. The correlations of the derivative objects exhibit a specific idealized experiment catching the most general and salient features of experimental practice.

A system of crossbred objects belongs to the derivative subsystems of both mature cross-theories. Consequently, the occurrence of inconsistent statements in the crossbred theory is identical to the synchronous occurrence of inconsistent statements in both mature cross-theories. The latter is identical to endowing the derivative objects with new properties, incompatible with the old ones.

It was already pointed out that a system of derivative objects is an invariant content of a certain class of empirical schemes. That is *why the occurrence of the cross-contradiction should lead* not only to the contradiction between the crossbred theory and a single experiment but *to the contradiction* of the crossbred theory *with the whole class of experiments.*

For instance, the occurrence of the cross-contradiction between mechanics and electrodynamics in the abovementioned crossbred theory - the planetary atom theory - is identical to endowing theoretical objects of electrodynamics with new properties incompatible with the old ones. This entails the following. The crossbred theory contradicts the experiments not only with certain types of atoms - with alkaline metals, for instance, - but the experiments with *all* the elements of Mendeleev's motley table as well. According to the predictions of the startling planetary atom theory, *all* atoms should be unstable.

The occurrence of the cross-contradiction in the theory of blackbody radiation entails the contradiction of the crossbred theory not only with the renowned experiments of Lummer & Pringsheim or Rubens & Kurlbaum. The infinity of the blackbody radiation energy density led to the odd situation, when

> "*all* the bodies should continuously radiate the electromagnetic waves, and not only outside, but inside also, until *all* the energy of the bodies is ultimately transformed into electromagnetic field energy and the temperature falls to absolute zero" (Landau 1958, p.81; my italics).]

Herewith a cross-contradiction manifests itself not by common deviations of theoretical predictions from experimental results. It is a commonplace that no theory ever managed to withdraw *all* its anomalies. When the problem situation is set forth by the cross-contradiction, such odd anomalies transpire that are *out* of the domain of usual deviations common for the mature theory under consideration. Namely these 'abnormal' anomalies may attract the attention of the scholars first and the period of ineradicable 'crisis' may commence.

The abnormal anomalies differ considerably from the common ones not only in their magnitude. Incidentally it was not rare in the history of physics when the anomalies were deftly withdrawn by judicial *non ad hoc* modifications of the partial theories. Yet in the particular cases of abnormal ('screwy') anomalies, the inability of the mature cross-theories to end with such anomalies constitutes a vital factor for the occurrence of the problem situation. Why a common anomaly cannot be considered as a deadly rebuttal of a theory, forsooth? - It is because such a theoretical object can always be inserted into a derivative theoretical system that can change the relations between other theoretical objects. Nevertheless, if the anomaly is not an ordinary one but is generated by the cross-contradiction, the situation changes gradually. At least two diverse mature theories have to grapple now with an exasperating anomaly.

In this intricate case, a reconcilement of one mature cross-theory with the observation data will inevitably lead to strong deviations from the same data of the other mature cross-theory ('the swing case'). As a result, the mature cross-theories can be brought to an agreement with the experimental results only after the cross-contradiction winding up, i.e. first after the successful accomplishment of the global theory creation.

In general, in social-psychological terms

> "[Kuhnian] crisis arises at the point when anomaly cannot be localized, when scientists sense that they no longer quite know how to go on becomes more general. Crisis is debilitating precisely because it cannot be localized as the falsification of some beliefs; it is instead the ***incoherence*** of research practice. The alternative to normal scientific practice being 'in the truth' is not error, but *confusion*" (Rouse 2013, p.5).

For instance, the phlogiston chemistry decisively failed as a practical tool for further exploring and coping with the relentless proliferation of sundry 'airs.' For the working 'normal scientist,' an inability to proceed with research in a coherent way must eventually be an inseparable objection to any theoretical program.

The theory-change model brings some novel light on rather stale dispute concerning the role of crucial experiments. According to the down-to-earth empiricist standpoint, the smashing single '*experimentum crucis*' is quite sufficient to refute a scientific theory. Yet according to the elevated pantheoretical doctrine, critical experiments for mature theories do not exist at all. The empiricist vulnerable conception is grounded on the following fallacious tenet: there exists the fairy observation language completely independent of theoretical propositions. On the other hand, the sublime pantheoretical conception exaggerates the theory-laidenness of observations. Hence, it is no wonder that, as stodgy partisans of extreme standpoints, the empiricists and the pan-theorists are right only partly. According to lucid theory-change model, there can be *any* number of critical experiments but only in the problem situation set forth by the cross-contradiction. An anomaly, engendered by the cross-contradiction, constitutes a smashing 'crucial experiment.' However, it is crucial not in the renowned 'falsificationist' sense. It cannot refute finally but it can limit the domain of validity only. If the experimental data are apparently out of the crossbred-theory predictions domain, it indicates that to eliminate the deviations one has to put up the global theory first. The common method of intruding some auxiliary modifications in any of mature cross-theories would be useless.

It is clear now why certain well-known experiments were regarded as 'crucial' only with hindsight, only some years *after* they were performed. It is unnecessary to ask what experiments - of Rubens & Kurlbaum or Lummer & Pringsheim – did actually refute the classical electromagnetic theory of radiation. In reality, both experiments were 'crucial,' as well as many other ones that have been forgotten up to now.

Besides the already mentioned, a long list of real crucial experiments embraces Young's famous experiment, abnormal Mercury perihelion precession observations, and so on. Yet the renowned Michelson-Morley experiment is not an indubitably crucial one. This typical 'bland crucial experiment' surely helped to choose between the partial and humble theories of Fresnel and Stokes vividly debating about partial or complete ether drag. Both mature theories were judiciously proposed within the same Lorentz- Langevin -Wien reductionist program.

Let me delve into the last example. The taken for granted argument of the empiricists is Lord Kelvin's prophetic speech, ardently devoted to Two Clouds 'that obscure the beauty and clearness of the dynamical theory.'

However a more thorough analysis elicits that the celebrated protagonist of the Classical Physics spoke not on 'hard facts' and 'ultimate refutations' but on two exasperating *paradoxes*. Kelvin punctiliously took the experimental evidence when illustrating the paradoxes only.

For instance, the First Paradox ('Cloud') – '*Relative Motion of Ether and Ponderable Bodies*' - was dealt with even by Fresnel and Young. It inextricably involved the intricate question, 'how could the earth move through an elastic solid, such as essentially is the luminiferous ether?' (Kelvin 1901, p.1). This paradox consists in the exasperating question: if 'ether does occupy the same space as ponderable matter' and if 'that ether is displaced by ponderable bodies moving through space occupied by ether,' 'how then could matter act on the ether?' (Kelvin 1901, p.3). The notorious Michelson & Morley experiment is envisaged only as making dubious one of the solution of the paradox (that of Fresnel), and the highly controversial 'ad hoc' contraction hypothesis of Lorentz & Fitzgerald is delightfully called 'brilliant' (Kelvin 1901, p.6).

No wonder that the renowned Second Cloud presents nothing else than the famous Maxwell-Boltzmann doctrine regarding the partition of energy. It had been characterized 'the greatest difficulty yet encountered by the molecular theory' even by James Clerk Maxwell in 1873. And the textbook blackbody experiments of Lummer & Pringsheim and Rubens & Kurlbaum are not even mentioned.

The occurrence of the abnormal problem situation (or of stern 'crisis' in Kuhnian terms) leads to gradual changes in the activities of scientific community. It strives to exchange the 'old' mature theory for new ones. The scholars are invigorated by the promising ideas that will later be reconstructed by the historians & philosophers of science as synthetic and reductionist 'hard cores' of the corresponding 'research programs.' As a rule, they are not fully aware that the sequences of theories they propose belong to strict and definite Procrustean forms of reductionist or synthetic programs. Furthermore, actually they are often unaware that they are putting the global theory up.

II.6. How a Global Theory is Put Up.

Is it possible to contrive the omnipotent global theory? Is the probability that scientists can stumble against the global theory in their chaotic wanderings significantly more than zero? Can the close-knit scientific community be compared to a flock of monkeys vividly inserting the whimsical texts on their antediluvian laptops? Is the probability to arrive at the global theory equal to the probability that the monkeys will print '*The Decameron*'? Or the '*Tractatus Logico-Philosophicus*'?

Certainly, a synthetic global theory is groped in reality by trial and error. Quite contrary to a reductionist global theory (with stiff and definite rules of construction), a synthetic mature theory is contrived gradually. In the course of the construction, intermediate hypotheses are discreetly proposed that happen to possess obvious opportunities of providing constant empirically progressive problem shift in comparison with reductionist programs. Let me exhibit them.

A basic system of theoretical objects and various systems of derivative objects as well possess two indissoluble assets (Stepin 2005). Firstly, a system of theoretical objects, a theoretical model is a sui generis *representation* (see Wartofsky 1979) of an array of objects under consideration. For instance, the solid basis of classical mechanics meticulously depicts mechanical motions with the help of the ideal model. The model immaculately represents the motions of real bodies by the motions of the 'material points' in the 'inertial systems of reference' under the influence of the 'forces.'

On the other hand, a system of theoretical objects makes up a model that seeks to portray the features of actual *experimental device*. This property permitted Vyacheslav S. Stepin to call a basic system 'A Fundamental Theoretical Scheme' (FTS) and a derivative system 'A Partial Theoretic Scheme' (PTS). For instance, the FTS of classical mechanics that represents mechanical motion by the motion of a material point (in an inertial frame of reference under the influence of the forces) features a *gedankenexperiment.* The latter catches the hallmarks of multifarious mechanical experiments with real bodies. The secure basis of Newtonian mechanics is a generalization of practical operations with real bodies moving along the inclined planes, operations with oscillating pendulums, with collisions of billiard balls, and so on. In particular, the PTS of small-oscillations constitutes a model describing the various features of the experiments with oscillations of real pendulums, real stretched strings, etc.

Just imagine now that the reductionist program has been successfully realized in some domain. It means that the global theory was contrived in such a way that the place of the Global Basis was steadily occupied by the basis of one of mature cross-theories. The basis of the second mature cross-theory eventually became the subordinate system of its derivative objects. This means that the same system of theoretical objects turned out to imbibe the motley features of rather *diverse* measuring operations. Their difference is caused by the fact that the basic systems of both mature cross-theories had been contrived *independently* of each other and long *before* they met. They were created as distinctive generalizations of independent and quite different operations.

For example, successful implementation of the Lorentz-Langevin-Wien eminent reductionist program (reduction of electrodynamics to mechanics) would portent the following. One of the derivative subsystems of Maxwellian electrodynamics, consisting of 'differentially small current' and 'magnetic field of a current,' – simultaneously catches the substantial features of the experiments with the bodies moving along the inclined planes, of the experiments with the rotations of rigid bodies, etc.

It is as plain as a pikestaff that the whimsical paradox springs out of the fact that the basis of one of mature cross-theories is probably unable to generalize the features of rather different measuring operations. The paradox can be swept away, and it was - in Lorentz's fine research program, for instance, - by artful contriving the 'operationally indefinite' object. It helps to 'wash away' the operational foundations of the 'fundamental' theory to such an extent that it becomes able to accumulate the measuring operations of the 'phenomenological' theory too. The peculiar ideal object – the notorious 'aether' - was inserted into the reductionist program of Lorentz, Langevin, and Wien as a 'carrier of electromagnetic oscillations.' It is a vivid example of dubious operationally indefinite object. No physical experiment could determine the motion through such 'aether' (see Einstein's 'Paradise' letter to Mileva cited in the following chapter).

The whimsical construct was inserted into the system of abstract objects of Lorentz's theory surely not due to certain generalizations of measuring operations. It was inserted for the construction of the mechanical basis from that of Maxwellian electrodynamics. The material points (the 'particles') were to be taken as 'ether perturbations,' whereas the forces acting on them were to be tentatively determined via the tensions of the ether.

The price of such a peculiar procedure may turn out to be rather high. It can lead to considerable decrease in the *predictive power* of mature theory. What is the cause of this power, forsooth? If a mature theory is *not* a direct generalization of empirical data, how can it predict the results of new experiments that are not even performed? How can a mature theory forecast future?

According to Stepin (2005), this opportunity is grounded on the stiff connections between the theoretical basis (the FTS) and real experiments, between theoretical and experimental practices. The indispensable link of the basic ideal model with the experiments is always *intermediated* by the systems of derivative objects. A basic object catches the characteristic features of the relations of the derivative objects that belong to the lower level of the organization of theoretical objects. Therefore to render a true operational definition of a basic object is to posit an *idealized* measurement operation, a gedankenexperiment, yet not the real one.

For example, the operational definition of electric field density **J** is provided not by picturing the real experiments with electrometers and charged balls. It is procured by the sedulous description of subtle interrelations of the Maxwell theory abstract objects - ‘electric field at a point’ and ‘test charge.’ These abstract objects are inherent components of the partial theoretical scheme of Maxwellian electrodynamics.

But *their own* operational status is determined now by real, and not by ideal, measuring operations. Viz., the ‘test charge’ is determined through such action of a massive charged body on the other, when the reciprocal influence of the second body on the first can be neglected.

The important bond between the PTS and ES levels is installed because all the PTSs constitute the idealized schemes of real interactions observed in *actual* experiments. *All* the bonds between *all* the levels of a mature theory should be stiff ones. This rigidity allows one to connect a hypothetical prediction referring to the upper level with all the underlying levels of a mature theory. Consequently, the insertion of operationally vague abstract objects, common for reductionist theories, should lead to a considerable *decrease* of the predictive power. It does not mean, of course, that a reductionist program cannot predict at all (recall the mechanical interpretation of heat, for instance).

Radical elimination of the cross-contradictions is chiefly possible in such global theory that tentatively contains all the cross-basises as *different* systems of derivative objects. In my humble opinion, only in synthetic global theory can the global theoretical scheme constitute a model that aptly combines all the salient features of idealized experiments with the abstract objects of *both* cross-theories. I had previously pointed out that the creation of the global synthetic theory is accompanied by the constant empirically progressive problem shift.

It is possible to explain now why synthetic programs *should* provide a larger empirically progressive problem shift than the reductionist ones. The reason consists in the successful utilization of the crossbred objects by the synthetic programs and their strict prohibition in the reductionist ones. Although the crossbred objects should be thrown with disgrace out of mature cross-theories, they are bestowed with the honourable rank of *global* theoretical objects in synthetic programs instead.

Indeed, according to their abovementioned definition, global objects are the same objects from which the basises of T_1 and T_1 are tentatively constructed. The global objects should necessarily contain information about $\{\mathbf{B_1}\}$ and $\{\mathbf{B_2}\}$, but the crossbred objects possess this hallmark too! They are tentatively constructed from $\{\mathbf{B_1}\}$ and $\{\mathbf{B_2}\}$. Furthermore, since the construction of the crossbreeds from the basises is possible, the reverse process of reconstructing basises from the crossbreeds should be possible also. I have already pointed out that in general, it is possible to construct *several* systems of crossbred objects.

A crossbred system contains only *part* of the information about their basises. Wherein only the whole set of various cross-systems possesses *all* the information about their random 'parents.' This set is always open for new components since it is impossible to declare beforehand that all the cross-domains are known to us. A mature theory can suffer unrestrained advancement by the construction of PTSs from the FTS for each novel domain of validity. That is why the global system may be ascertained only via careful amalgamation of all the crossystems and cannot live a secluded life. It is always open for new crossbred systems; moreover, the introduction of novel elements can endow the old ones with novel promising properties.

This general scheme can be illustrated by the example of the transition from XIX- th century physics to the physics of the XX-th century. Viz., the modern relativistic quantum theory was created by the consequent generalization of Bohr's crossbred atom models, Einstein's crossbred semi-corpuscular radiation theory, STR, Dirac's relativistic electron theory and so on. The process is still not finished up to the present (see Wigner's keen 1958 comments).

Coordination of new-fangled STR with the celebrated Newton's theory of gravity ultimately led to the General Theory of Relativity (GTR) contrivance (see chapter V and Nugayev's 2018 monograph). Einstein's strenuous efforts to incorporate Newtonian gravity into the STR peculiar framework commenced already in 1907 when he had to write a detailed review for '*Jahrbuch der Radioaktivitat*' (Einstein 1907; see Vizgin 1981 for details). His primary effort turned out to be surprisingly unsuccessful since simple-minded (though lorentz-invariant) generalizations failed to explain properly the anomalous perihelion motion of Mercury.

But let us return to the general scheme of global theory creation. Why can the dazzling synthetic theories empirically supersede the jogtrot reductionist ones? *The insertion of the crossbreeds engenders contradictions in all mature cross-theories and forces them to accommodate each other, to 'interpenetrate' and to 'enrich' each other.*

For instance, let us first turn to significant example with the pivotal theories of XIX-th century physics. According to such a shrewd witness at the events as Max Planck, the leader of German school of theoretical physics,

> "In fact, the impulse for an extension of mechanics had to *come from without*, and it came from the electro-dynamical theory" (Planck [1908], 1960, p.83; my italics).

Furthermore, Planck amends the example, furnishing the following further jolly important additions:

> "To-day [viz., circa 1908], it is generally accepted that an absolutely rigid mechanical theory of ether consistent with Maxwell's simple equations cannot be countenanced and this makes a *gap between classical mechanics and electro-dynamics that cannot be bridged*. It only remains to define exactly the range of applicability of the two theories *or to modify one of them*. The former soon proved to be impracticable, for mechanics and electro-dynamics overlapped on the question of motion of electrons and the manner of settling the controversies was shown

> for the first time by the discovery of deviations from the laws of classical mechanics, the deviations finding expression in the variable inert mass of the electron. *Einstein's theory of relativity contains a simple, complete solution of the problem of reconciling mechanics and electro-dynamics* in a quite general way, as far as it retains the practical essentials of the classical theory and still fulfils the demands of electro-dynamic calculations. The modification in mechanics due to the principle of relativity is the introduction of a new universal constant, entirely foreign to classical mechanics, namely, the velocity of light in an absolute vacuum" (Planck [1908], 1960, p.85; my italics).

However, not only STR turns out an unexpected result of the mature theories' interpenetration. The other apt example is the old quantum theory itself (Planck [1908], 1960, p.85).

Secondly, the dynamic of GTR creation had been continually governed by internal tensions between two research traditions, that of STR and Newton's theory of gravitation. The encounter of the traditions, their profound interpenetration, and subtle intertwinement entailed coming into being the rival programs of construction of the relativistic theory of gravity. This is about the programs of Abraham and Nordström – on the one hand – and Einstein's deft program – on the other. It should be punctuated that the intricate relations between the GTR and its artful rivals were far more complicated in 1907-1915 than it may seem from the pestered 'truth-falsity' dilemma. Thus, one can once more contemplate with surprise the interlacement and *interpenetration* of rival 'incommensurable paradigms' into each other.

The relentless encounter rendered contrivance of the hybrid domain at first with an irregular set of theoretical models. The rival approaches of Einstein, Abraham, and Nordström grew closer and closer; and eventually, Einstein transformed them into complementary 'mathematical' and 'physical' strategies of a systematic research program that engendered the notorious 'Entwurf' and then the GTR. Systematically, on tentative eliminating the contradictions between the models constructed, the hybrid set was put into order by Einstein via 'the principle of equivalence.' Likewise, as in the STR startling case, Einstein was able to freely juxtapose Nordström, Abraham's, and his non-metric theoretical schemes without reducing ones to the others.

Hence, it can be exhibited (see Chapter V for details; see also Nugayev 2018) that the main reason for the GTR 'victory' over the rival programs of Abraham and Nordström was a synthetic asset of Einstein's program. This farsighted program did supersede the rival ones because it masterfully assimilated some substantial ideas of the Nordström sober program (Nordström 1912, 1913a, b), as well as some felicitous premises of the judicious program of Abraham (1912a, b, c, d; 1915). Viz.,

(i) Guileless invention of a relativistic theory of gravity had dawned with the crossbred object construction in Einstein's 1907 farsighted paper. The crossbred object insertion – the discreet introduction of inertial and simultaneously gravitational mass ($E = mc^2$) – led *perforce* to STR methods intrusion into the Newtonian theory of gravity and to reverse penetration of Newtonian gravity methods into the STR. As a result, both mature theories were radically restructured and the corresponding changes in both of them were set up. The substantial changes were epitomized in the particular sequences of crossbred models, the byproducts of the transformation performed.

(a) On the one hand, an inevitable consequence of the STR invasion into the Newtonian theory of gravity turned out to be Nordström and Abraham's distinctive scientific research programs.

(b) On the other hand, no less relentless, owing to pervasive 'equivalence principle,' was the Newtonian theory invasion into the STR. It inevitably led to the sequence of Einstein's conspicuous papers on the generalization of the relativity principle and to spreading the principle not only on inertial systems of reference but on the various accelerated systems as well (Einstein 1912a, b; 1913 a, b, c; 1914, 1915). Nevertheless, the most valuable result of the hybrid theories of Nordström and Abraham consisted in that both theories contained *very promising hints* on how the global theory could be put up. Hence, the climax of the stage was Einstein's daring proposal of thorough 'equivalence principle' that became one of the firm GTR heuristic foundations.

(**ii**) The basic GTR model was sedulously constructed due to the proper amalgamation of the hybrid models of Einstein, Nordström and Abraham invented within diverse research programs. The asset of Einstein's dazzling style of reasoning was his open-mindedness. His renowned anti-dogmatism made it possible for Einstein to freely juxtapose Nordström, Abraham's, and his non-metric theoretical schemes without blunt reducing one to other.

(**iii**) It is this asset that helps to comprehend the true reasons for Einstein's victory over the rival programs of Nordström and Abraham. The 'Entwurf' and then the GTR artfully 'incorporated' Nordstrom and Abraham's deft accounts in a higher-level theory.

Thus, a true reason for Einstein's victory over the rival programs does not by no means consist in that he peremptorily rebutted the alternative accounts. On the contrary, it consists in that he subsumed them all into a novel mellow and grand framework. Hence, the resolute drive in advancing the GTR came from the craving attempts to match physical and mathematical approaches, embodied in the Abraham, Nordström, and Einstein's crossbred theoretical models. A trustworthy explanation of Mercury's perihelion motion turned out to be a bewitching by-product of strenuous reconciliation efforts.

(**iv**)The thought-provoking volumes (Renn et al. 2007a) are permeated with the idea that GTR "was not just a one-man affair and a lucky individual discovery but the result of a profound transformation of the extended system of knowledge of classical physics." According to the authors, the GTR genesis can be taken as an example of a sheer 'Copernican process' that is repeatedly observable at decisive moments in the history of science.

"This process implies that marginal elements of a complex architecture of Knowledge torn asunder by inner tensions become the starting points for reconstruction as a result of which a new structure emerges that is, however, essentially *composed of the building blocks already available*. This process may be designated as a 'Copernican process'" (Renn and Sauer 2007, p.47; my italics).

In a sense, our reconstruction (see also Nugayev 2018) can be regarded as an attempt to *explicate* and *furnish* the aforementioned doctrine of the

'Copernicus process,' rendering a *mechanism* of simultaneous resolute transformation of 'old' classical paradigms. The hallmark of the mechanism is the construction of the crossbred systems from the basic theoretical objects of 'old' mature theories. These crossbred systems are subsequently generalized to erect a structure of a novel mature theory. A mature physical theory can be empirically successful if it represents a kind of 'ein entwurf' (a draft) that can be explicated by the results of new experiments that increase its empirical content. In this way, GTR could become such a helpful vehicle for the production of new gravitational propositions. It managed to reconcile and intertwine the 'old' classical traditions of Newtonian mechanics (including Newtonian gravity) and STR in such a masterful way that they ceased to contradict each other. And their newly contrived fusion could not only successfully explain, but also fortuitously predict novel experimental data embracing relativistic astrophysics and cosmology.

Thirdly, in the modern GTR &QFT startling case (examined in detail in Nugayev 2020), "the general theory of relativity and the quantum theory of fields, *taken together in a new synthesis*, have thus far *enriched* each other" (De Witt 1980, p.683; my italics). The enrichment manifested itself first and foremost in the downright penetration of QFT into GTR (the gravitational field quantization). On the other hand, much less known reciprocal GTR penetration into QFT showed itself in the discovery of thrilling 'Unruh *effect*' in flat space-time.

> "By building a simple model of a particle detector, and accelerating it in the vacuum state in flat space-time, it is possible to show that such a detector behaves exactly as if it were immersed in a thermal sea of particles of temperature T = 8πa/kc" (Unruh 1980, p.154).

Although the first efforts to reconcile GTR and QFT started from the very moment of creation of GTR (theories of Weyl, Cartan, Kaluza & Klein), the substantive progress was attained quite recently. At least at the first stages, it surprisingly seems to have nothing to do with superstrings, supergravity, and so on. I reckon that the real progress in QFT and GTR unification commenced with the hybrid objects' construction in a brand new crossbred realm called the '*Quantum Field Theory in Curved Spaces.*' (Birrell & Davies had issued in 1981 the first review under the same title). The crossbred realm embraces at least two important domains – (BP) black hole physics and (IC) inflationary cosmology. Let me examine the examples in more detail.

The Black Holes. The most apparent example of genuine cross-theory is Hawking's path-breaking 1975 paper on the particle creation by black holes. Yet the real story started earlier, with Jacob Beckenstein's pioneering inquire into the black hole entropy S. His 1973 paper turned out a masterpiece artfully combining deep physical intuition with outrageous flies of imagination so rare in the dark waters of shabby math speculations. Beckenstein's salutary quest had commenced with a bewitching analogy between entropy and the surface of the black hole horizon. From the fecund information theory standpoint, entropy and horizon of a black hole are the quantities of the same kind. If a particle manages to get under the horizon of a black hole, the surface of the hole increases due to the mass addition. At the same time, entropy increases also, since one cannot look under the horizon to know the state of the particle there. Consequently, the area of the horizon should be equated to entropy with some proportionality coefficient.

J.M. Bardeen, B. Carter, and S.W. Hawking (1973) thoroughly generalized and clarified Beckenstein's startling, yet semi – intuitive, ideas. They elicited the profound analogy between black hole physics and thermodynamics via more strict and mathematically sophisticated language. The fine expressions were sedulously derived for the mass of a stationary axisymmetric solution of the Einstein equations containing a black hole surrounded by matter and for the difference in mass between two solutions.

> "Two of the quantities which appear in these expressions, namely the area A of the event horizon and the 'surface gravity' k of the black hole, have a *close analogy* with entropy and temperature respectively. This *analogy suggests* the formulation of four laws of black hole mechanics which correspond to and in some ways transcend the four laws of thermodynamics" (Bardeen, Carter, Hawking 1973, p.161; my italics).

Now GTR and thermodynamics were merged into a general theoretical scheme, and the corresponding conspicuous results showed up rather quickly. According to the Second Generalized Law of Thermodynamics, formulated by Beckenstein and Hawking, the flow of entropy across the event horizon of a black hole should increase the area of the horizon A. The important quantity "entropy + some multiple of A" never decreases. Yet if one equates the surface gravity of a black hole k to temperature T, a startling paradox occurs. If a black hole is immersed in the sea of blackbody radiation at a *lower* temperature (than that of the black hole), the Second Law will be necessarily violated! Note that the only way out to avoid the paradox is to admit that a black hole should emit particles at a steady rate with a temperature equal to $k/2\pi M$. Namely, that committed Stephen Hawking (1975) by shrewdly considering quantum-field effects. At least several important lessons can be learned from the story. Firstly, three mature theories were involved in the occurrence of the cross-contradiction: thermodynamics, GTR, and QFT. Besides, QFT is laid out as a necessary link connecting GTR with thermodynamics. Without QFT, GTR cannot be properly reconciled with thermodynamics and statistical mechanics. The conspicuous story looks like that: classical thermodynamics 'perceives' something about the quanta, as if the quantum theory in some curious way is 'contained' in the thermodynamics. In the light of the already told, this conclusion does not seem jolly strange; the quantum theory already is a wayward fusion of thermodynamics, mechanics, and electrodynamics.

> "This has perhaps contributed to the idea that gravity has a somehow *deeper* relationship to thermodynamics than to other branches of physics" (Gould 1987, p. 449; my italics).

From the very beginning, the QFT in Curved Spaces met with a throng of paradoxes caused by the lack of proper agreement between GTR and QFT (Fulling 1973; De Witt 1975). The guileless notion of a 'particle' unexpectedly turned out to be observer-dependent. Furthermore, exasperating conformal anomalies suddenly occurred; various renormalization methods furnished diverse results for vacuum energies; the queer negative energy fluxes began to fill the space with GTR, and many other such bizarre contraptions began to bother poor theorists and continue to puzzle them. It is important for the developed model that at least one of the leading black-hole boffins had professed: the cause of all these paradoxes consists in the semiclassical nature of black-hole evaporation theory.

> "The gravitational field is treated 'classically' (as a space-time metric) there, while its sources are described in terms of quantum field theory. The basic equations of GTR, $G_{\mu\nu}= T_{\mu\nu}$, contain *incompatible* quantities on the two sides of the equals sign. Furthermore, this incompatibility has consequences even at the present experimental scale" (Unruh 1980, p.154; my italics).

One of the most fascinating crossbreeds here is the whimsical QFT vacuum. When it was intruced to the GTR derivative system (the black hole physics), it did acquire a new and rather queer property – 'to produce infinitely great gravitational field.' However, the common methods of its regularization (the so-called 'cut-off procedures') were elaborated for flat spacetimes only. Whereupon they are useless in strong gravitational fields.

The introduction of the QFT vacuum restructured the relations between the GTR objects significantly, and the process of the global theory construction inevitably began. Besides the black holes, various inflation scenarios are the indispensable stages of the global theory construction too. The enticing creation of the synthetic global theory seems to follow the promising lines of Einstein and Bohr.

Inflationary Cosmology. The whimsical QFT vacuum gained particular urgency with the advent of quantum theory.It was realized that all fields of Nature are endowed with irreducible 'zero-point energies,' even in their vacuum state. Note that the notion of zero-point energy is a pure result of quantum theory and has no classical counterpart. Incidentally, it was introduced by Planck in an address to the German Physical Society in February (Planck 1911).

Grounded on the ideas underlying his second theory (Kuhn 1978), Planck had punctiliously calculated the average energy of an oscillator vibrating with frequency ν and with absolute temperature T:

$E = h\nu/1 + h\nu/\exp (h\nu/kT) -1.$

The value of the quantized energy levels of an oscillator could be written as

$$E = (1/2)\,(E_n - E_{n-1}) = (n+1/2)\,h\nu,$$

where n= 0, 1, 2,…At T= 0 the average energy is not zero but equals the finite energy (1/2) hν.

For a long time, Planck's bold hypothesis was considered dubious and controversial. The point of view that zero-point vacuum energies and fluctuations are indeed real was first evidenced by the so-called Casimir effect going back to 1948. Casimir predicted from the QFT basis that there exists an attractive force between two perfectly conducting parallel plates

$$F \approx d^{-4},$$

where d is the distance between the plates. According to Casimir, the difference between the zero-point energies in a pure vacuum and the presence of the plates would be measurable and it was (Sparnaay 1958).

Initially, the Casimir effect was not taken relevant for cosmology. For a long time the cosmological constant λ, famously introduced in 1917 by Einstein to keep the Universe stable, and the quantum mechanics of the vacuum lived separate lives. The situation only changed in the late 1960s. Inspired by a revival of interest in cosmological models with a positive cosmological constant, Yakov B. Zel'dovich (1968) stressed the significance of λ in the context of QFT. He also elicited the first version of what would be known as a cosmological-constant problem.

> "With Zel'dovich's work two historical strands were finally joined; the quantum vacuum and the energy density related to the cosmological constant" (Kragh and Overduin 2014, p.57).

Zel'dovich's influential works initiated the contrivance of another important crossbred realm – the inflationary cosmology. Inflation is the idea that the very early Universe may have been dominated by vacuum energy, giving rise to a short period of vastly accelerated expansion. However, inflation did not obtain wide recognition until high-energy particle physicists became involved. Stimulated by the advances of electroweak unification, they started to examine the implications of spontaneous symmetry breaking for cosmology, with the attendant possibility of a false vacuum state. Also, up to now the most compelling argument for inflation consists in its potential to link the observed large-scale structure in the present-day Universe to antediluvian quantum fluctuations in the earliest moments after the big bang.

Yet ultimately, despite an apparent success of inflation models in addressing many conundrums of the early Universe, theoretical cosmologists were reluctant to entertain the idea that vacuum energy might also dominate the Universe later until they were obliged to do so by astronomers. Improvements in detector technology, image analysis, and other factors led to the definitive detection of dark energy in 1998 using the magnitude-redshift relation for Ia supernovae. The quick convergence of two independent teams to a similar conclusion played a large part in convincing the cosmological community of the reality of a vacuum-dominated accelerating Universe. Yet we do not know why λ is so much smaller than calculations based on QFT would suggest (Perlmutter 1999).

Thus, vacuum plays a pivotal role in the entire modern crossbred realm, being a steady global object of the future theory: 'making everything out of nothing' (Robert Weingard) during the process of spontaneous symmetry breaking. Incidentally, the Quantum Vacuum played the leads at each stage of the unification of the four fundamental forces of nature - in Higgs mechanism establishment (Higgs vacuum), in fifth coordinate compactification in Kaluza-Klein models (the Casimir effect), in the Green & Schwartz superstring model of Quantum Gravity free of gravitational anomalies.

The budding encounter of T_1 and T_2 increases their domains of validity on D_1 and D_2 respectively. The domain of validity of T_1 expands owing to D_2, and vice versa. The domain of validity of both mature cross-theories becomes equal to two (D_1+D_2). For instance, the promising creation of foolhardy photon theory (and of startling STR) was due to intense interpenetrations of the following mature theories.

(i) Of Boltzmann's statistical mechanics and classical thermodynamics (Willard Gibbs's and Albert Einstein's pioneering inquiries into the realm of statistical thermodynamics).

(ii) Of Boltzmann's statistical mechanics and Maxwellian electrodynamics (Einstein's 1904 punctilious papers on fluctuations of the electromagnetic field).

(iii) Of classical thermodynamics and Maxwellian electrodynamics (Max Planck pioneering introduction of the notions of temperature and entropy of the blackbody radiation).

(iv) Of Newtonian mechanics and Maxwellian electrodynamics (STR basic postulates, retrieval of corpuscular theory of light).

The advancement of Einstein's ideas by Louis De Broglie (see Pais 1980, for details) consisted in endowing the 'ponderous matter' with wave properties, etc.

Thus, crossbred objects constitute the queer 'channels' through which the ideas and methods of mature cross-theories interpenetrate. Let us compare synthetic theories with reductionist ones, which, as was already mentioned, strictly forbid the construction of the crossbreeds. Behold the most favourable for reduction case, viz., when the fundamental problems are successfully solved already, and the domain of validity of the global theory becomes larger since D_2 was successfully added to D_1. The 'fundamental' theory T_1 has penetrated the 'phenomenological' theory T_2, but the reverse process is nevertheless prohibited. Therefore the domain of validity of the synthetic ideal is two times larger than the domain of validity of the reductionist one. Namely that hallmark may provide the ultimate victory of synthetic programs over their reductionist rivals. Although each global theory should withdraw the cross-contradictions according to the model proposed, the *real* history of physics is more intricate. The process of global theory contrivance may last for centuries (Copernican revolution) embracing new cross-domains with the cross-contradictions playing the lead of powerful 'driving forces' in the creation of new mature theories.

For instance, in his defiant "*De Revolutionibus*," a queer Catholic canon Nicolas Copernicus had audaciously pioneered (waywardly following Aristarchus of Samos) in taking the Earth as a common planet orbiting the Sun. Thus he had contrived a crossbred theoretical model capable of descending the mathematical principles from divine phenomena to the mundane ones (further implemented by Galileo). On the other hand, through the same crossbred object, the principles of Earth physics were genuinely ascended (Newton) from mundane objects to the Skies (see Nugayev 2013 and 2022 for details).

Likewise, in intricate '*Treatise on Electricity and Magnetism*,' a pre-eminent Compendium, a whimsical Cambridge post-graduate James Clerk Maxwell had contrived a queer crossbred object – the displacement current. It enabled him to extrapolate the electromagnetic principles on the optical phenomena, and vice versa. His Copernican deeds consisted now in strengthening arguments for the identification of electromagnetic and luminiferous ethers and the invention of the *crossbred theory* mixing Faraday's field concepts with that of the Ampere-Weber electrodynamics. It was the displacement current that allowed artful amalgamation the theories of electromagnetism and optics (Nugayev 2015).

In the case of relentless clash between GTR and QFT, the developed scheme can be animated by the following quotation.

> "It is ironic that just at the time when the experimental tests are becoming possible, are being performed, and are verifying the predictions of the theory, the most determined assault on the structure of the theory itself is taking place. In the attempt to make general relativity compatible with the quantum theory of fields, many feel that the theory must be *altered* in some fashion. The most notable example of this is, of course, supergravity, with its addition of the gravitino" (Unruh 1980, p.153; my italics).

Furthermore, the most remarkable hallmark of the modern supergravity theories consists in that all of them were punctiliously constructed without experimental results. Yet

> "the second reply to such criticisms is that we have already a powerful 'experimental' fact that needs to be explained, and that is the *logical inconsistency* of the two theories" (Unruh 1980, p.154; my italics).

Note that when contriving modern global theory, the experimental data are taken into account *indirectly*, being artfully embedded into pieces of theory. The lead of contradictions between Theory and Experiment is still played by the cross-contradictions between several theories.

Be that as it may, the true 'hard fact' consists in that the contrivance of the Global Theory had already begun. Indecisive results including newfangled supergravity were punctiliously obtained without *any* experimental results. This is not to buttress the complacent standpoint since too many steps are to be taken until one can allege that the subtle unification machine is already at hand.

What remains crucial in establishing the supergravity is its empirical justification. In my humble view, the latter should consist first in establishing the *operational status* of the concepts involved. This status can be (partly) achieved by ideal measurement procedures that are analogous to the pre-eminent Bohr-Rosenfeld procedures in quantum electrodynamics. As is well-known, to render an operational definition of a physical quantity, connected with a basic object, is to procure a description of idealized measurement procedures.

> "It was only by *asking questions about measurements*, about how the various phenomena would *affect measuring instruments* such as real-model gravity wave detectors or observers falling through the 'singularity' at R=2M of the Schwarzchild metric, that these issues were resolved" (Unruh 1976, p.128; my italics).

It is no wonder that promising results were obtained namely in this direction of research. I have in mind the startling Unruh effect:

> "The lesson of general relativity was that the proper questions to ask of a theory are those that concern the behaviour of detectors that have been fashioned to detect the effects that concern one. For example, in trying to decide whether particles are created, one needs to define a particle detector within the theory; i.e. some model of a physical system that will respond to the presence of a particle. In my attempt to apply *this philosophy* in the mid 1970s, I looked at the behaviour of a simple model of a particle detector (1) for some quantum field, which, for simplicity, I took as a massless scalar field F" (Unruh 1976, p.128; my italics).

Eventually, the generality of the methodological scheme outlined is most strongly buttressed by an example rather far from physics the so-called '*Devonian controversy*' (Rudwick 1985, 2017). The controversy erupted in the 1830s among leading practitioners of the science of 'geology.' It was concerned with the classification and nomenclature of certain major formations (sets of strata) in relation to others. Yet it was seen to challenge the dominant exemplar-embodied in the practice of stratigraphy – that formations could be identified, and correlated between one region or another, by finding them everywhere. They could then be arranged unambiguously in a unique structural order.

The notorious 'Devon controversy' arose when the abovementioned well-established practice of stratigraphy was extended from the relatively easy cases of the younger formations to the more subtle cases of much older and more disturbed rocks. And the controversy was triggered when highly respected English geologist Henry De la Beche discovered fossils characteristic of the Coal formation, which was of supreme economic importance in the early Industrial Revolution, in the far older strata then recently named Cambrian. This anomaly was so radical and so unexpected – and its potential economic significance so important – that the factual reliability of the report was immediately questioned by other pundits, notably Roderick Murchison and Charles Lyell himself.

It should be stressed that the detailed historical evidence

> "invalidates *any* claim that the resolution of the problem signalled the 'triumph' of one side of the initial argument and the 'defeat' of the other [...] Instead, it shows how the *social process of controversy*, with all sides deploying the changing empirical evidence to their best advantage, repeatedly forced the actors to *modify* their positions. Out of this social process, a third and eventually successful alternative emerged, which *had not been foreseen* by either side at the outset; it incorporated elements derived from ***both*** the initial rivals [i.e. the crossbred objects], yet it was no mess compromise (see Fig. 12.1). In other words, the consensual solution to the focal problem resulted in the production of genuinely new knowledge [...] its consensual product is the defining of a distinctive 'Devonian period' in the history of the Earth, during which, for example, *both* plant and animal life made their first significant appearances on land. Was this new knowledge a social construction or a discovery about the world? It was, of course, *both*" (Rudwick 2017, p. 184; my italics).

CHAPTER III.

Maxwell's Masterful Unification of Optics and Electromagnetism and the Incommensurability Tenet.

"Kuhn's original and quite modest concept of paradigms – as concrete pieces of research that act as exemplars for fruitful future work – remains, I think, much more useful for understanding the making of scientific knowledge than his later concept of paradigms as alternatives that are radically incompatible and incommensurable. Fruitful exemplars have often emerged from a ***social*** *process of controversy, entailing both conflict and collaboration, within a limited 'core-set' [Harry Collins] of active researchers"*
— **Martin Rudwick**

Abstract. It is contended that one of the promising directions for solving the problem of incommensurability of paradigms, coined by T. Kuhn and P. Feyerabend, can be associated with the trend of advancement of neo-Kantian epistemology, embodied by the writings of Ernst Cassirer. According to Cassirer, the statements fixing connections and relationships between mathematical ideal constructs form a 'neutral language' that can serve as a firm ground for comparing the 'old' and 'new' paradigms.

A case study of the genesis and functioning of a neutral mathematical language related to the Maxwellian synthesis of optics and electromagnetism is provided. It is elicited that its basis is constituted by the mathematical language of continuum mechanics. The main function of the neutral language was to project meticulously the consequences of all the unified theories onto the mathematical model, 'rewrite' all known laws in this mathematical language, compare their conclusions with each other to eliminate contradictions and generalize the stuff in a self-consistent system of Maxwell's equations.

Keywords: incommensurability, neutral language, Cassirer, mathematatization, Maxwell, synthesis.

III.1. Introduction.

Typifying the post-positivist philosophy of science, Thomas Kuhn and Paul Feyerabend had formulated and substantiated the eminent thesis on the incommensurability of paradigms replacing each other in the scientific revolutions (the 'incommensurability tenet'). Accordingly, the creators of new mature scientific theories (Newton, Maxwell, Einstein) make such profound transformations of theoretical ontologies, so resolutely reject obsolete ideas that the 'new' paradigms they construct become incommensurable with the 'old' ones. This means that in science there are no *objective* bases for comparing successive paradigms.

Thus, in his epoch-making "*The Structure of Scientific Revolutions*," T. Kuhn provides a classic example of incommensurability related to the process of replacing Newtonian mechanics with special relativity:

> "But the physical referents of these Einsteinian concepts are by no means identical with those of the Newtonian concepts that bear the same meaning. (Newtonian mass is conserved; Einsteinian is convertible with energy. Only at low relative velocities may the two be measured in the same way, and even then they must not be conceived to be the same [...] The normal-scientific tradition that emerges from a scientific revolution is not only incompatible but often actually incommensurable with that which has gone before" (Kuhn [1962], 1970, pp. 102-103).

Paul Feyerabend ([1975], 2010) adhered to a similar, although less rigid and more structured position The so-called 'incommensurability problem' consists in that if the 'new' and 'old' paradigms are really incommensurable, then on what grounds can we argue that the 'new' paradigm is better than the 'old' one? And if we cannot assert this, then what is the progress in the development of science?

Moreover, since the 'old' and 'new' paradigms are radically separated from one another by the impenetrable barrier of 'incommensurability,' it is difficult to see how one gets from one to the other.

> "Kuhn faces a severe difficulty explaining how new ideas emerge against the background of their predecessors, from which they are so totally distinct" (Renn 1993, p.312).
>
> Forsooth,
>
> "the new paradigm, or a sufficient hint to permit later articulation, emerges all at once, sometimes in the middle of the night, in the mind of a man deeply immersed in crisis. What the nature of that final stage is – how an individual invents (or finds he has invented) a new way of giving order to date now all assembled – must have remain inscrutable and may be permanently so" (Kuhn [1962], 1970, pp.89-90).
>
> Furthermore,
>
> "There are many things to be said about the nature of progress in the sciences; the thing that you *cannot* I think say coherently is that they get closer and closer to the truth… [Hence] That's what you do within a system, - judge the truth or falsity of statements. Across a system you can't apply that sort of calculation… [Thus] I take theories to be whole systems, and as such they don't need to be true or false" (Thomas Kuhn's interview to Skuli Sigurdsson 2017, p.24).

However, it is not accidental that in view of one of outstanding figures in modern history& philosophy of science,

> "Kuhn's original and quite modest concept of paradigms – as concrete pieces of research that act as *exemplars* for fruitful future work – remains, I think, much more useful for understanding the making of scientific knowledge than his later concept of paradigms as alternatives that are *radically incompatible* and *incommensurable*. Fruitful exemplars have often emerged from a ***social*** process of controversy, entailing *both* conflict and

collaboration, within a limited 'core-set' [Harry Collins] of active researchers" (Rudwick 2017, p. 182, my italics).

Therefore, of undoubted interest for today philosophy of science are those accounts that precede the study of specific problems of the philosophy of science by the adoption of a general view of the evolution of science and, accordingly, clearer criteria for scientific progress. This, as a rule, is possible when this view is formulated as a continuation of one or another *classical* philosophical tradition. It makes possible to apply the well-structured apparatus of classical epistemology, which, in particular, makes it possible to distinguish 'truth' from 'error' more clearly. Such a promising standpoint is the epistemology of Ernst Cassirer, advanced in line with the neo-Kantian philosophical tradition. The latter, as part of the ongoing rejection of neo-positivist concepts, is attracting more and more attention (Ryckman 2005; Ladyman 2008, et al). The purpose of this chapter is to analyze the concept of Cassirer from the point of view of the problem of (in) commensurability (first and second sections) and to reveal, using the vivid example of the Maxwellian revolution, what possibilities it provides for solving this problem (third section).

III.2. Neo-Cantian Epistemology of Ernst Cassirer.

In accordance with the traditions of the Marburg school of neo-Kantianism, the basic concepts of each science, the means by which it raises its 'ultimate' questions and formulates certain conclusions, are not passive reflections of human existence, but 'intellectual symbols' constructed by people themselves (Cassirer 1923). And it was physical and mathematical cognition that most clearly realized the symbolic nature of its fundamental means of knowledge before the others. According to Cassirer, modern science arose only in the XVth - XVIIth centuries. Only when two diverse research traditions that had previously developed relatively independently of each other met and began to 'interpenetrate and actively interact' in Western European culture.

The first tradition (Bacon) is the tradition of experimental natural science, according to which all scientific knowledge arises out of experience. The second tradition (Euclid) is the tradition of mathematically immaculate reasoning, according to which every scientific statement must be proven, i.e. in logically impeccable way derived from the system of self-evident axioms.

Interaction of these traditions constituted one of those antinomies, the inevitability of which continues to develop natural science. The construction of systems of ideal mathematical objects constitutes the result of how

> "the fundamental forms of knowledge penetrate into each other - sensory sensation and pure contemplation, categories of pure reason and ideas of pure reason; how the theoretical image of reality is determined by their interconnection and interaction" [Cassirer 1923, p.14].

Thus, the laws of classical (Newtonian) mechanics make sense only when they describe connections and relationships between 'ideal ultimate images,' which we abstractly put in place of empirical sense data of sense perception. Accordingly, mechanical movement is a predicate that cannot be directly applied to the objects of the sensory world around us. It makes sense

> "only when applied to the class of objects that mathematics substitutes in their place in its free creativity" [Cassirer 1910, p.143].

Being radically opposed to the traditional logical teaching of Aristotle, the path of mathematical formation of concepts was blazed by Galileo, Kepler and Newton 'through the method of forming series.' The latter does not at all consist in eliciting from a certain throng of homogeneous impressions what is common to them, but in establishing

> "a certain principle by virtue of which the different follows one from the other. The unity of the concept was found not in some unchanging sum of attributes, but in the rule, due to which a simple difference was presented as a regular series of elements" (Cassirer 1910, p.172).

As a result, in place of single 'values of the series,' established by observation or measurement, one puts "limiting values towards which the series as a whole move." Natural science theories provide illustrative examples of the application of this constructive method of 'passing to the limit': a perfectly elastic body, an ideal gas, an incompressible fluid, a perfect circular process, etc. Classical mechanics itself describes the strict laws of motion only through the use of the ultimate idea of a 'material point.'

But, in introducing this 'realm of ultimate ideal objects,' we must in no way confuse it with the world of Platonic ideas or forms described by Kepler:

> "Mathematical ideas are the eternal prototypes and 'archetypes' according to which the Divine Builder of the world arranged everything."

As Cassirer himself never tired of repeating,

> "it is no longer possible to be mistaken and take the objects of physics - mass, force, atom, ether - as new *realities* that must be explored and inside of which one must penetrate, once we understand that they are [merely] *tools* created by thought to depict the chaos of phenomena in the form of a dismembered and measurable whole" (Cassirer 1910, p.193, my italics).

Using numerous examples described in detail in the works of the great German physicist Heinrich Hertz, Cassirer concludes that the fundamental concepts of the natural sciences, and especially the basic concepts of mechanics and electrodynamics, are by no means *copies* of things directly given in the experience. On the contrary, they are the necessary 'constructive blueprints' of physical thought. The theoretical significance of these projects is primarily determined by the extent to which the consequences derived from them coincide with the experimental results obtained. On the other hand, any concept of natural science, in the specific meaning that it has in mathematics and physics, is

> "merely a *symbol* for a certain form of connection, which has thrown off one by one any special material content. It means only the kind of possible coordination, but not the '*essence*' (das Was)

of the elements that must be coordinated with each other" (Cassirer 1910, p.236, my italics).

As a result, instead of conjuring up a new truly Platonic world behind the world of perceptions, our knowledge limits itself to sketching out the general logical schemes with which it expects to represent the connections and relations of perceptions. A natural-science symbol has its correlate not in the constituent parts of perception, but in the regular *connection* that exists between its parts.

III.3. Concept of Scientific Progress in the Epistemology of Ernst Cassirer.

In accordance with the entrenched Kantian tradition, Cassirerian epistemology proceeds from the fact that modern science categorically rejects natural philosophical claims to penetrate 'inside' nature, implying its last substantial foundations. The development of science appears in the Cassirer version of neo-Kantian epistemology as follows

I. First, a careful differentiation is made of the individual spheres of perception offered to us by empirical contemplation; in each of them, 'fluid transitions' are replaced by precise, well-fixed numerical definitions. At the same time, the 'world of physics' (however, like the 'worlds' of chemistry, biology, etc.) is subdivided and articulated according to differences directly given in sensation. Optics corresponds to light and color sensations, thermodynamics corresponds to thermal sensations, and acoustics corresponds to sound sensations. But already at this stage, in order to enter the theoretical schemes of optics, thermodynamics, etc., the sensory content must go through the transformations described above, when certain sensations are replaced in the primary schemes by certain limits of mathematical sequences or, in Russian vocabulary (Stepin 2006) "particular theoretical schemes."

II. But inevitably the second, more significant stage comes.

"The genuine and much more difficult task of physics is not to pass from sensible 'qualities' to precisely defined quantities. Rather, after the implementation of this preparatory work, she

> raises the really fundamental question for her about *the functional connection* and connection of individual qualitative areas. They should not be made out in simple coexistence and succession, but they should be thought of as a law-like and law-governed unity" (Cassirer 1923, p.340).

To illustrate the thesis, Cassirer refers to Max Planck's famous report "*The Unity of the Physical Picture of the World.*" The founder of quantum physics convincingly substantiates in it the tenet according to which true progress in the natural sciences consists in ever greater *liberation* of theoretical thought from the restrictions imposed on it by the initial attachment to the direct content of sensory sensations and to the primitive way of their articulation, i.e. in a decisive 'emancipation from anthropomorphic elements.' In particular, the advancement of physics and chemistry did not stop at a purely empirical establishment of the existence of various constants, but moved on to the establishment of general systemic relationships that made the appearance of certain complexes 'understandable.'

This 'comprehensibility' essentially meant the search for some universal laws that determine the connections of heterogeneous constants. With the help of these laws, areas that were previously taken extremely distant from each other began to be considered from a single point of view. At first, this approach was limited to certain areas, but then it became the 'main trend of modern physics.' First, the constants describing physical and chemical processes approached each other. But a closer convergence of the constants occurred only in the middle of the XIXth century. It happened when James Maxwell revealed the relationship between the constants that determine the optical properties of certain substances and the constants that characterize their electrical conductivity.

At the same time, the 'general schematism of the concept of number' plays a decisive role in the evolution of the second stage. Namely:

> "Number functions as a kind of *abstract medium* in which different sensory areas meet each other, losing their sensory heterogeneity in the process. For example, in Maxwell's theory, the phenomenon of light is identified with the phenomenon of

electricity, since they receive the same numerical value and expression. The external difference between optical and electrical phenomena disappears as soon as it is recognized that the constant c, appearing in Maxwell's equations, is exactly equal to the speed of light in vacuum. It is the form of a purely numerical relation that helps to overcome the heterogeneity of sensory properties and achieve the homogeneity of the physical 'essence'" [Cassirer 1923, p.346].

Note that the leader of German theoretical physics at the beginning of the XXth century, cited above, came, in the course of reflection on the foundations of his professional activity, to similar conclusions (Planck 1922). Returning to the Maxwellian synthesis, one can conclude that, as its most important result, it was recognized that all the laws valid for light rays (refraction, interference, polarization, etc.) are equally applicable to thermal radiation. Thus, a strong 'union' was ascertained between these two areas, which made it possible to overcome the qualitative difference in sensations in which warmth and light are given to us. From now on, they began to differ - in the 'objective sense' of physical judgments - only by numerical values, only by a certain parameter denoting the 'wavelength' of the two varieties of electromagnetic radiation.

Eventually

"the consistent development of a genuinely physical form of thinking and its symbolism gave rise to a [neutral!] language that created the possibility of such an unheard-of expansion, which disclosed that the boundaries of sensation are merely random 'anthropomorphic' ones" (Cassirer 1923, p. 347).

But the most apparently considered feature of the second stage evolution was manifested in the history of the discovery of the law of conservation of energy in Robert Mayer's formulation. Cassirer emphasized that for the latter the mechanical law of conservation of active force means nothing more than

"a universal relation connecting various areas of physical phenomena and making them *qualitatively comparable and*

> ***commensurable*** [...] The principle establishes fixed numerical relations in the transformation of heat into motion and motion into heat, but it does not in any way surmise that heat in its essence is nothing else, as movement" (Cassirer 1923, p.353).

Eventually, as a result of the joint successes of Maxwell, Lorentz and Einstein, the theory of the electromagnetic field and the special theory of relativity were created. Thence the ether, as 'wooden iron,' amalgamating mutually exclusive features, was discarded, and completely reigned such a reality as the 'physical field.' But at the same time

> "the reality we call 'field' is no longer conceivable as a complex of physical 'things,' but it is expressible as a set of physical relations" (Cassirer 1923, p. 365).

The 'essence' of light no longer consists for us in the 'movement' of a wave or in 'oscillations,' but only in the periodic changes of a vector, whose direction is always considered to be perpendicular to the direction of the light. Progress in the development of physics does not at all consist in a deeper immersion into the essence of the processes under consideration, expressed in discovering, finally, the true 'substance' from which all things are made. It consists in combining various physical theories, in bringing the integrity of natural phenomena under one universal rule that establishes connections and relationships between various sections of physics (Nugayev 2020).

And the result of this unification should be a new system of relations describing the connections between already known idealizations of physical processes. Thus, the process of unification of scientific theories turns out to be inextricably linked with another deeper process expressing the progress of scientific knowledge. As a result of the unification of physical theories

> "With the individual peculiarity of impressions, its inner heterogeneity has also disappeared, so that regions that, from the point of view of sensation, are *completely incomparable* with each other, can now be understood as interconnected members of the same general plan. Only in this lies the special value of scientific

constructive building; in it turns out to be connected by means of continuously stretching intermediate logical terms that which in the first naive view is alien and unrelated to each other" (Cassirer 1923, p.352, my italics).

III.4. A Case Study of Neutral Language Contrivance and Functioning: Maxwell's Masterful Synthes of Optics and Electromagnetism.

It is a commonplace that Maxwellian electrodynamics was a stage of the Faraday program advancement grounded on the field concept (see, for instance, Chalmers, 1975 and references cited therein). The latter had provided prediction and corroboration of startling electromagnetic waves phenomenon and have superseded at last the Ampère-Weber research program based on the action at a distance concept.

However a more pervasive account of the XIX-th century physics that came into being due to the studies of Daniel Siegel (1991), Margaret Morrison (2000) and Olivier Darrigol (2001) enables one to challenge this point of view as an oversimplification – and to proffer a modified version of it – with a help of the following reasons.

(i) At first, James Maxwell himself many times, beginning from his first paper and up to the last one, had stressed that the pivotal ideas of the Ampère- Weber electrodynamics were *complimentary* and *not* alternative to those of the field concept. Even starting his electrodynamics studies, on May 1855, a post-graduate student at Cambridge wrote to his father

"I am working away at electricity again, and have been working my way into the views of heavy German writers. It takes a long time to reduce to order all the notions one gets *from* these men, but I hope to see *my way* through the subject, and arrive at something intelligible in the way of a theory" (quoted from Campbell l& Garnett, 1882, p.105).

And, at last, in his epoch-making "*Treatise on Electricity and Magnetism*" Maxwell pictures the contrivance of his system of equations as follows:

> "I was aware that there was supposed to be a difference between Faraday's way of conceiving phenomena and that of the mathematicians, so that neither he nor they were satisfied with each other's language. I had also the conviction that this discrepancy did not arise from either party being wrong" (Maxwell [1873a], p.499).

(ii) No-one at eminent Cavendish laboratory - conducted and well-equipped by Maxwell - had made a serious and sustained attempt to confirm Maxwell's theory (Mahon, 2003). Maxwellian electrodynamics' main verification was provided only by a star student of Hermann Helmholtz whose name was Heinrich Hertz (1888). And the latter did not regard himself a follower of Maxwell (Darrigol, 2001).

(iii) Faraday's well-known influence on Maxwell was strongly exaggerated. The explanation of the field concept adoption due to the sympathy to intermediate action is not confirmed by Maxwell's papers more thorough analysis (Shapiro, 1973).

It reveals that Maxwell started to take the field notion as an inherent part of physical reality sufficiently late: only *after* he had derived the electromagnetic waves existence from his equations, i.e. *after* the derivation of the 'displacement current.' Up to that moment he had used the field notion only as an illustrative means for building up blunt pictorial images of complicated vector differential equations.

For instance, within one and the same paper Maxwell (1856) uses different models for illustrating different analytic relations, punctuating that:

> "It is not even a hypothetical fluid which is introduced to explain actual phenomena. It is merely a collection of *imaginary* properties which may be employed for establishing certain theorems in pure mathematics in a way more intelligible to many

> minds and more applicable to physical problems than that in which algebraic symbols alone are used" (Maxwell [1856], 1890, p.160).

The aim of the present scrap is to take a further step in providing such a rational reconstruction of Maxwellian electrodynamics genesis that takes into account (i) – (iii). The reconstruction should provide a 'theoretically progressive problemshift' relative to other reconstructions contending that Maxwellian revolution is a more complex phenomenon than appears from the standpoints of some well-known scientific revolution models (Kuhn, 1977; Lakatos, 1978).

I had already underscored that one of these models' drawbacks consists in that they lack the mechanisms of the paradigms' (or scientific research program's *interactions* -Nugayev, 1999; 2012). And it is one of my pivotal aims to exhibit that the remark is especially appropriate for Maxwellian electrodynamics genesis. I'll try to indicate that the Maxwellian program had superseded that of Ampère-Weber namely since it had constantly communicated with it. The Maxwellian program did embrace some of the propositions of the Ampère-Weber 'hard core,' as well as some surmises of the Faraday and Young-Fresnel programs.

In my view one of the main drawbacks of the preceding studies consists in *underestimation of Maxwell's own methodology* deployed for his ambitious project of mechanics, electrodynamics and optics unification. Though in every domain of creativity (including epistemology) Maxwell always took his own way. And he tried to teach his students in the same way too. Recall the following passage of his Marichal college speech:

> "It is best that every man should be settled in *his own* mind, and not be led into other men's ways of thinking under the pretense of studying science" (quoted from Mahon, 2003, p.70).

As the author of ground-breaking "*Treatise on Electricity and Magnetism*" himself put it in one of his letters, "I find I get fonder of metaphysics and less of calculations continually" (quoted from Campbell & Garnett, 1882, p. 298).

One should not forget to recall Gustav Kirchhoff: "He is a genius, but one has to check his calculations."

In my view one should take Ludwig Boltzmann's comments on Maxwell's works more seriously. In immaculate lectures on Maxwell's theory as well as in his comments on Maxwell's electrodynamics papers (that he had translated into German), the founder of statistical mechanics had stressed that many Maxwell's works but especially his *early electrical papers* "were not properly understood." Probably it can be explained by the fact that these works "written according to the *long-term plan*" disclose that their author "was *as mastermind in theory of knowledge as he was in the field of theoretical physics*" (Boltzmann, 1895). Maxwell's genuine neo-Kantian epistemology that sprung out from an intention to find a fruitful compromise between the extremes of Kantian relativism and Scottish 'common sense realism' was an indispensable part of his ambitious unification of optics and electromagnetism design.

Withal, surely Maxwell was not the first to unify optics and electromagnetism. Yet he did not approve the way his predecessors had done it. Why?

> "The present state of electrical science seems peculiarly unfavorable to speculation...No electrical theory can now be put forth, unless it shows the *connexion* not only between electricity at rest and current electricity, but between the attractions and inductive effects of electricity in both states...The results of this simplification may take the form of a purely mathematical formula or of a physical hypothesis. In the first case we entirely lose sight of the phenomena to be explained; and though we may trace out the consequences of given laws, we can never obtain more extended views of the *connexions* of the subject" (Maxwell, [1856], p.155).

His eminent predecessors were Hans Christian Oersted (1777 – 1851), Andre-Marie Ampère (1775-1836), Wilhelm Weber (1804-1890), Michael Faraday (1791-1867) and William Thomson (1824-1907). Yet Maxwell's Weltanschauung, as well as his epistemology, was characterized by *an extraordinary high level of philosophical culture.* A brilliant student at Edinburgh and Cambridge and a post-graduate at Cambridge was enchanted by a profound skepticism of David Hume, George Berkley and especially Immanuel Kant imbibed at the lectures of Sir William Hamilton on mental philosophy at Edinburgh University.

For instance, in the 03.25.1854 frank letter Maxwell states that

> "I have been reading Berkeley on "The Theory of Vision," and greatly admire it, *as I do all his other non-mathematical works*; but I was disappointed to find that he had at last fallen into the snare of his own paradoxes" (quoted from Campbell& Garnett, 1882, p.109, my italics).

Likewise,

> "Comte has good ideas about method, but no notion of what is meant by a person" (Campbell &Garnett, 1882, p.1o8).

Hamilton's dazzling lectures, which were a prominent component of the Scottish university curriculum, 'interested him greatly.' From the Class of Metaphysics his mind 'gained many lasting impressions' (Lewis & Campbell, 1882). The conspicuous lectures of Sir Hamilton made a lasting impression on him, in "stimulating the love of speculation to which his mind was [too] prone" (Niven, 1890)

Luckily, sir William Hamilton (1788-1856) was one of the most outstanding representatives of Scottish 'common sense philosophy,' a worthy heir of Thomas Reid and James Stewart. Yet it is no coincidence that he had visited Germany two times to study Kantian philosophy. He indefatigably stressed *Kant's proposition that all knowledge is relative*; so we know nothing about things themselves except by their *relationship* to other things. Correspondingly, he had stimulated a spirit of criticism in his pupils by insisting on the great importance of psychology as opposed to the older stale metaphysical method.

Hamilton's 'philosophy of the conditioned' surely had an unequivocal Kantian flavor. Like Kant, he held that one can have stable knowledge only of "the relative manifestations of an existence, which in itself it is our highest wisdom to recognize as beyond the reach of philosophy." Though unlike Kant *he had argued for the position of a 'natural realism'* in the Reidian tradition.

The Reverend Thomas Reid (1710-1796) directed his thought-provoking "*An Inquiry into the Human Mind on the Principles of Common Sense*" [1764] against Hume and Berkeley. It is here – he asserted – that the 'danger of the idealism' lies – in its reduction of reality to "particular perceptions," essentially *unconnected* with each other. The independent unit of knowledge is not an isolated impression but a judgement. Yet in such a judgement is contained the reference both to a permanent subject and to a permanent world of thought, and, implied in these, such judgements, for example, as those of existence, substance, cause and effect. The sublime principles are not derived from sensations, but are 'suggested' on occasion of sensation, in such a way as to constitute the necessary conditions of having perceptive experience at all.

The sober doctrine of relativity of knowledge has seemed to many – including James Stewart Mill – contradictory to his realism. But for Hamilton, the two are firmly held together by a kind of intuitionism that emphasizes certain facts of consciousness that are both primitive and incomprehensible. They are though constitutive of knowledge, "less forms of cognitions than of beliefs" (quoted from Audi, 1999, p.360).

The subtle relativism or phenomenalism which Hamilton adopted from Kant and sought to engraft upon Scottish philosophy is absent from the original Scottish doctrine.

Resolutely denying Hume's skepticism, Hamilton did his best to *find a compromise* between Kant's relativism and Reid's realism. And it was namely the compromise that Maxwell have pointed out as a basic point of his metaphysical program on moving from Edinburgh to Cambridge:

> "in the meantime I have my usual superfluity of plans…4. Metaphysics – *Kant's Kritik of Pure reason in German, read with a determination to make it agree with Sir W. Hamilton*" (Campbell & Garnett, 1882, p.77).

The 'Copernican revolution' in epistemology famously initiated by Kant consisted in that the world of common every-day experience (i.e. Husserl's 'lebenswelt') had lost its dominating position in interpreting things that can be perceived by our senses. Kant had vigorously exchanged the world of common experience by the world of Galilean experimental and mathematical physics grounded on the idealizations of the 'lebenswelt' phenomena. Hence truth became not only something spontaneously revealing and disclosing itself but something that can be comprehended only by a special ('scientific') method.

On the other hand, if truth is comprehended only in experience and one can grasp not 'the things by themselves' but just the 'phenomena,' it is necessary to decidedly reject the very opportunity of reaching the absolute truth. Our sensory representation is by no means a representation of things 'in themselves,' but merely of the way in which they appear to us. Hence the so-called '*analogies of experience*' are especially important in Kant's epistemology.

> "This type of cognition is cognition *according to analogy*, which surely does not signify, as the word is usually taken, an imperfect similarity between two things, but rather a perfect similarity between two relations in wholly dissimilar things. Such is an analogy between the legal relation of human actions and the mechanical relation of moving forces: I can never do anything to another without giving him a right to do the same to me under the same condition; just as a body cannot act on another body with its motive force without thereby causing the other body to react just as much on it. Right and motive force are here completely dissimilar things, but in their relation there nonetheless complete similarity. By means of such an analogy I can therefore provide a concept of a relation to things that are absolutely unknown to me" (Kant [1783], 2002, p.146-147).

In more detail, in '*The Critique of Pure Reason*,' Kant elucidates a more interesting example:

> "Prior to the perception, however, and therefore completely a priori, we are able to cognize its existence, provided it stands in connection with some perceptions according to the principles of the empirical conjunction of these, that is, in conformity with the analogies of perception. For, in this case, the existence of the

> supposed things is connected with our perception in a possible experience, and we are able, with the guidance of these analogies, to reason in the series of possible perceptions from a thing which we do really perceive to the thing we do not perceive. Thus, *we cognize the existence of a magnetic matter penetrating all bodies from the perception of the attraction of the steel-fillings by the magnet,* although the constitution of our organs renders an immediate perception of this matter impossible for us" (Kant [1787], 2010, p.170).

Note that *even the example of the analogy of experience was taken by Kant from the domain of electromagnetism* thereby, as if showing the way to Maxwell. The latter had punctuated many times that things we can measure directly, like mechanical force, are merely the outward manifestations of deeper processes, involving entities like electric field strength, which are beyond our power of visualization (see Mahon, 2003).

A more detailed exposition of Maxwell's research program that he had followed through all his life is provided by his following truly philosophical works – a speech "*Are There Real Analogies in Nature?*" read at the 'Apostles' Cambridge club in 1856 (just after the publication of his thorough paper "*On Faraday's Lines of Force*," 1855-1856) – and his stupendous paper "*Helmholtz*" (1877).

Discreet Cambridge speech is not a crude exposition of Kant's epistemology but a tense discussion of Maxwell with 'Kant in himself.' It is not accidental that its very heading contains a question and not a proposition "*Are There Real Analogies in Nature*?" - Maxwell did not give the final answer - in full accord with Kant's antinomies that represent human Reason's futile attempts to overstep the limits of Experience. He multiplies arguments pro and contra the strong assertion that there certainly *are* real analogies in Nature.

He commences the speech contending that

> "the whole framework of science, up to the very pinnacle of philosophy, seems sometimes a dissected model of nature, and sometimes a natural growth of the inner surface of the mind" (Campbell& Garnett, 1882, p.133).

Or in the other part of the speech:

> "are we to conclude that these various departments of nature in which analogous laws exist have a real interdependence; or that their relation is only apparent and owing to the necessary conditions of human thought?" (Campbell& Garnett, 1882, p. 134).

Certainly Maxwell's thinking in terms of Kantian antinomies is only a facet of his approach. In Hamilton's wake, Maxwell tries to find a *way* of *his own* between the Scylla of Kantian transcendentalism and the Charybdis of Scotch common sense realism.

In modern times the Scottish view of knowledge is characterized by the following principles (Mertz, 1964; Olson, 1975).

1. All knowledge is relational.
2. Analogies (and models connected with them) are among the chief such relational models of knowing.
3. Analogies and models are necessary for psychological reasons. For common people, understanding requires the use of models for simplifying and organizing knowledge.
4. Strong psychological tendencies in the Scottish Common Sense tradition admit reconciliation with logical and analytical components of Kantian philosophy.

Hence for Maxwell the effective resolution of the antinomies should consist in adopting *partial points of view*, as 'all human knowledge is partial.' What remains is establishing correspondences or analogies.

> "Whenever they see a relation between two things they know well, and think they see there must be a similar relation between things less known, they reason from the one to another. This supposes that although pairs of things may differ widely from each other, the relation in the one pair may be the same as that in the other. Now, as in a scientific point of view the *relation* is the most important thing to know, a knowledge of the one thing leads us a long way toward a knowledge of the other. If all that we know is relation, and if all the relations of one pair of things correspond to

> those of another pair, it will be difficult to distinguish the one pair from the other, although not presenting a single point of resemblance, unless we have some difference of relation to something else whereby to distinguish them. Such mistakes can hardly occur except in mathematical and physical analogies…Perhaps the 'book,' as it has been called, of nature is regularly paged; if so, no doubt the introductory parts will explain those that follow, and the methods taught in the first chapters will be taken for granted and used as illustrations in the more advanced parts of the course; but if it not a 'book' at all, but a magazine, nothing is more foolish to suppose that one part can throw light on another" (Maxwell; quoted from Campbell & Garnett 1882, p. 124).

That is the first lesson properly taught by Kantian epistemology – (I) "*the principle of relativity of scientific truth.*" Note that even the examples of the analogies are taken by Maxwell from the '*Prolegomena.*' Hence it is not surprising that the next principle – (II) 'theory laidenness of observation' – is also extracted from Kant:

> "The dimmed outlines of phenomenal things all *merge into one another* unless we put on the focusing glass of theory, and screw it up sometimes to one pitch of definition and sometimes to another, so as to see down into different depths through the great millstone of the world" (Maxwell; quoted from Campbell & Garnett 1882, p. 125).

The importance of the last principle for Maxwell's methodology cannot be overestimated. In nature all the phenomena are tightly *interconnected* and merge into one another.

All the differences in theoretical approaches are due to the fact that their authors focus on the multifarious facets and different levels of the phenomena investigated. Hence a theoretician's task is to provide the special notions to cover the various domains of experience. But where should he discover them? – In experience, via immediate generalizations of the experimental data? – Another piece of Maxwell's creativity – his keen 1854 letter – makes it possible to dig still deeper into his thought laboratory:

> "It is hard work *grinding out* 'appropriate ideas,' as Whewell calls them. I think they are coming out at last, and by dint of knocking them against all the facts and half-digested theories afloat, I hope to bring them to shape, after which I hope to understand something more about inductive philosophy that I do at present.
>
> I have a project of *sifting* the theory of light and making everything stand upon definite experiments and definite assumptions, so that things may not be supposed to be assumptions when they are either definitions or experiments" (Maxwell; quoted from Campbell & Garnett, 1882, p.112).

Now it is clear where the 'appropriate ideas' come from: they are not the slavish copies of things, but are just the *a priori* forms by which a hotchpotch of sensations is "brought to order." According to Maxwell's distinctive essay "*Has everything beautiful in Art its original in Nature?*" (spring of 1854),

> "as the Theoretic and Imaginative faculty is far in advance of Reason, he can apprehend and artistically reproduce natural beauty of a higher order than his science can attain to" (Campbell& Garnett, 1882, p.133).

At first the 'appropriate ideas' are vague and dim; yet in the long run they are 'grinded out' by knocking them with the 'facts,' as well as other theories. Though theoretician's task is not only to introduce and furnish the subtle notions 'reflecting' the different facets of the phenomena under consideration, but also to unify the notions in effective synthesis.

The stages of such a synthesis are highlighted in Maxwell's chef-d-euvre "*Hermann Ludwig Ferdinand Helmholtz*" that begins as follows:

> "Hence the ordinary growth of human knowledge is by accumulation round a number of *distinct* centers. The time, however, must sooner or later arrive when two or more departments of knowledge can no longer remain independent of each other, but *must* be fused into a consistent whole. But though

men of science may be profoundly convinced of the necessity of such a fusion, the operation itself is a most arduous one. For though *the phenomena of nature are all consistent with each other*, we have to deal not only with these, but with the hypotheses which have been invented to systematize them; and it by no means follows that because one set of observers have labored with all sincerity to reduce to order one group of phenomena, the hypotheses which they have formed will be *consistent* with those by which a second set of observers have explained a different set of phenomena.

Each science may appear tolerably consistent within itself, but before they can be combined into one, each must be stripped of the daubing of untempered mortar by which its parts have been prematurely made to cohere.

Hence the operation of fusing two sciences into one generally involves much criticism of established methods, and the explosion of many pieces of fancied knowledge which may have been long held in scientific reputation" *(Maxwell [1877], 1890, p.592).

This passage is not accidental for Maxwell's creativity. In other writings he himself emphasized the asset of '*cross-fertilization of the sciences*' (Maxwell, [1873a], 1890, vol.2, p.744) evoking the image of bees pollinating flowers (see Harman 2003 for details).

The typical example of 'the daubing of untempered mortar elimination' principle (III) for Maxwell was

"the progress of science in Newton's time [which] consisted in getting rid of the celestial machinery with which generations of astronomers had encumbered the heavens, and thus 'sweeping cobwebs off the sky'" (Maxwell [1873b], 1890, p.315; see also Nugayev, 2012).

It should be added that the ideas of Maxwell's 1877 paper were generalized and elaborated by a whole number of scholars (Tisza, 1963; Podgoretzky & Smorodinsky, 1980; Nugayev 1985a;1999; 2018). As a result, a lucid 'mature theory-change model' was proposed. According to the model, the origins of scientific revolutions lie not in a clash of mature research traditions with facts, but of 'old' mature traditions with each other, leading to contradictions that can be successfully eliminated in a more general research program. Onwards, on the basis of the model the so-called '*conception of communicative rationality*' that tries to take the socio-psychological peculiarities of theory change into account was proposed. The doctrine enables one to describe the basic paradigms' interaction, as well as the interaction of cognitive and of social parts of the paradigms too. The key role in theory change is played by the proponents of the old paradigms' *dialogue* that leads to genuine interactions in all the three main dimensions including personal institutional (scientific community) and cultural (theoretical) aspects.

It seems to me that some of the features of the theory-change model and communicative rationality conception resemble the following stages of Maxwell's program realization.

First, "*A Treatise on Electricity and Magnetism*" (1873a) was mainly an encyclopedia and a textbook. The basic electromagnetic results were obtained in a sequence of three epoch-making papers: "*On Faraday's Lines of Force*" (1856), "*On Physical Lines of Force*" (1861) and "*A Dynamical Theory of Electromagnetic Field*" (1864).

The first one (Maxwell 1856) is dedicated to elaboration of the '*physical analogies*' method sprung out of Kantian epistemology. The method decidedly rejects the 'ontological' approaches striving for the 'essences' of electrical and magnetic phenomena and proclaiming that 'in reality' electricity and magnetism are sheer 'fields' and not 'action at a distance' phenomena, or vice versa. Maxwell's proposal is to take Faraday's lines of force as a kind of inscrutable tubes filled with ideal incompressible fluid.

> "I propose then, […]; and lastly to show how by an extension of these methods, and the introduction of another idea due to Faraday, the laws of the attractions and inductive actions of magnets and currents may be clearly conceived, without making assumptions as to the *physical nature* of electricity, or adding anything to that which has been already proved by experiment.

By referring everything to the purely geometrical idea of the motion of an imaginary fluid, I hope to attain generality and precision, and to avoid the dangers arising from a premature theory professing to explain the cause of the phenomena" (Maxwell [1856], 1890, p. 159).

It is crucial for a neo-Kantian that this incompressible fluid has nothing to do with experimentally given reality. The single constrain on the concocted models consists in the demand that the mathematical constructs should not contradict each other. In all the other delicate matters the physical analogies method admits an unlimited freedom of imagination. Even the relentless conservation laws can be broken down!

"There is nothing self-contradictory in the conception of these sources where the fluid is created, and sinks where it is annihilated. The properties of the fluid are at our disposal, we have made it incompressible, and now we suppose it produced from nothing at certain points and reduced to nothing at others" (Maxwell [1856], 1890, p. 162).

And in the other parts of the paper Maxwell exhibits the ways by which the idea of incompressible fluid motion can be applied to the sciences of statical electricity, permanent magnetism, magnetism of induction, and uniform galvanic currents. It should be stressed that the core element of his innovations consisted in the construction of '*neutral language*' for description and comparison of the consequences from the rival theories and competing theoretical models. Maxwell's 'neutral language' was not Carnap's and Reichenbach's 'observation language' directly springing out from the 'protokolsatze' generalizations. Maxwell is aware of the theory- laidenness of the observation data:

"experimental laws already established, which have generally been expressed in the *language of other hypotheses*" (Maxwell [1861], 1890, p.162).

He clearly comprehends that every observation nextricably carries the footprints of the theoretical language that helps to describe it. ('The daubing of untempered mortar,' as he will call them later in fascinating '*Helmholtz*' paper).

In order to compare and to integrate in a theoretical scheme lacking contradictions all the results of dissent experiments carrying the footprints of multifarious theoretical languages, it is necessary to construct an artificial theoretical language equally distant from the languages of theories under comparison. This stupendous language turned out to be the solid state mechanics (with hydrodynamics as its inherent part). Maxwell's ultimate aim was to rewrite all the known empirical and theoretical laws of electricity and magnetism applying the neutral language and then to compare them in order to contrive a system without contradictions. The final result of the 1856 magnificent paper was a system of equations lacking the eminent 'displacement current.'

It was not accidental that one of the main drawbacks of the incompressible fluid model consisted in that the latter, apart from some simple cases, was unable to explain interrelations and interactions of electrical and magnetic fields and electric currents, as well as Faraday's (1845) startling interconnection between optical and electromagnetic phenomena.

Though Maxwellian program's ultimate goal was to reveal the connection "between electricity at rest and current electricity" absent in the Ampère-Weber electrodynamics. Was it reached already in 1856? – Certainly not. The important connection between the current density **j** and the charge density ρ was lacking in Maxwell's initial 1856 scheme. It was to appear later, after the 'displacement current' insertion and finding out its consequence – the continuity equation $\operatorname{div} \mathbf{j} + \frac{\partial \rho}{\partial t} = 0$.

Hence, in 1861 the publication of the second paper (Maxwell 1861) consisting of four parts begins. Its aim was to rederive the results of Weber and Neumann deft theories on the basis of a new mechanical model grounded on the vortices of incompressible fluid concept.

> "My object in this paper is to clear the way for speculation in this direction, by investigating the mechanical results of certain states of tension and motion in a medium, and comparing these with the

> observed phenomena of magnetism and electricity" (Maxwell [1861], p. 162).

Again and again he has to emphasize that

> "the author of this method of representation does not attempt to explain the *origin* of the observed forces by the effects due to these strains in the elastic solid, but makes use of the mathematical *analogies* of the two problems to assist the *imagination* in the study of both" (Maxwell [1861], 1890, p.163; my italics).

The model sprung out of William Thomson's inquiries. He had revealed that the connection between magnetism and electricity has the same mathematical form as that between certain parts of the phenomena, of which one has a *linear* and the other a *rotatory* character. Note that that Thomson introduced the vortices model in incompressible fluid while investigating Faraday's experiments on the rotation of the plane of polarized light when transmitted along the lines of magnetic force. Ipso facto, they were the efforts to theoretical reconstruct the Faraday effect that provided the meeting of optics and theory of magnetism.

In the second Maxwell's model M_2 magnetic field was represented now by a set of vortices in incompressible fluid with the axes of rotation coinciding with the direction of magnetic field at a point.

> "Let us now suppose that the phenomena of magnetism depend on the existence of a tension in the direction of the lines of force, combined with a hydrostatic force [...] The explanation which most readily occurs to my mind is that the excess of pressure in the equatorial direction arises from the centrifugal force of vortices on eddies in the medium having their axes in directions parallel to the lines of force" (Maxwell 1861, p.165).

But now the important role of neutral language is played not by tube hydrodynamics but by a theory of stresses in the medium where vortices materially represent the vector operator 'curl,' while the necessary relations among the forces are described by mathematicians with the help of *tensors*. The general type of a tensor describing the general type of stress embraces combination of three principal pressures or tensions, in direction at right angles to each other. The tensor apparatus of solid state mechanics provided contrivance of a 'dialect' of new neutral language. It enabled to calculate the force acting on an element of the medium:

$\mathbf{F} = \mathbf{F}_1 + \mathbf{F}_2 + \mathbf{F}_3 + \mathbf{F}_4 + \mathbf{F}_5$.

(1) The first term $\mathbf{F}_1$ refers to the force acting on magnetic poles;

(2) the second term $\mathbf{F}_2$ refers to the action on bodies capable of magnetism by induction;

(3)-(4) the third $\mathbf{F}_3$ and the fourth $\mathbf{F}_4$ terms refer to the force acting on electric currents; (5) the fifth term $\mathbf{F}_5$ refers to the effect of simple pressure that lacks direct electromagnetic analogy.

Yet one of the most intricate problems of the vortices model (that puzzled even Daniel Bernoulli who invented it in XVIII century – see Whittaker 1910) was as follows. How can the rotation be transferred from one vortex to another so that "vortices in a medium exist side by side, revolving in the same direction about parallel axis"? – The artful conception that buttressed Maxwell in conceiving this kind of motion was that of the vortices being separated by a layer of particles called the "*idle wheels*." Is it possible to apply these particles for electricity?

And in the second part of his trailblazing 1861 paper "*The Theory of Molecular Vortices applied to Electric Currents*" Maxwell comes up to the *hardest problem* of his research program: what is "the physical connexion of these vortices with electric currents, while we are still in doubt as to the *nature of electricity*." It is this point where Maxwell has to trespass the principal *limits* of pure mechanical models and to *insert the alien elements of action at a distance theory*! Or, using the language of the abovementioned lucid methodological model, we can conclude that Maxwell had to construct the '*crossbred models*' from the languages of *both* cross-theories that combine the properties of completely different theoretical schemes.

According to Maxwell's queer model, an electric current is represented by the transference of the moveable particles interposed between the neighboring vortices. As a result,

> "these particles, in our theory, *play the part of electricity*. Their motion of translation constitute an electric current, their rotation

> serves to transmit the motion of the vortices from one part of the field to another, and the tangential pressures thus called into play constitute electromotive force. The conception of a particle having its motion connected with that of a vortex by perfect rolling contact may appear somewhat *awkward. I do not bring it forward as a mode of connexion existing in nature*" (Maxwell [1861], 1890, p. 345; my italics).

On introducing such abstract theoretical objects as 'electrical particles' and 'electric current representing the motion of such particles' Maxwell had significantly deviated from Faraday's adroit notions. According to Michael Faraday, the electrical charges should be taken as constructed by the ends of lines of force: they lack an independent substantial existence. Correspondingly, in his *genuine* research program the electric current has to be envisaged not as the motion of real particles but as an 'energy axis.'

This is the nub of the British field program: the fields are primary, while the particles are only secondary. Maxwell's blunt eclecticism was followed by H.A. Lorentz's subtle dualism. Lorentz initiated it in pervasive 1875 paper:

> "I shall start with instantaneous action at a distance: thus we will be able to found the theory on the most direct interpretation of observed facts" (quoted from Darrigol, 2001, p.323).

Thereby, it was not a temporary retreat. Even after 1861 Maxwell many times inserted the stiff notions of the Ampère-Weber atomism into his theories (Darrigol 2001).

Yet the results obtained were surely insufficient; the theoretical derivation of Coulomb's law was apparently lacking. Namely that was done in the third part of profound 1861 paper "*The Theory of Molecular Vortices applied to Statical Electricity*." Note that the vortices model contained too many ad hoc assumptions. We have now came close to the so-called '*Maxwell's miracle*.'

It turned out that if one (in the course of Fresnel optics and electromagnetism theory encounter) transposes the ether properties from optics to electromagnetism, she can eliminate at least one ad hoc supposition. Indeed,

> "it is necessary to suppose, in order to account for the transmission of rotation from the exterior to the interior parts of each cell, that the substance in the cells possesses *elasticity of figure*, similar in kind, though different in degree, to that of observed in solid bodies. The undulatory theory of light requires us to admit this kind of elasticity in the luminiferous medium, in order to account for transverse vibrations. We need then be surprised if the magneto-electric medium possesses the same properties" (Maxwell [1861], p. 13).

This peculiarity has a vital significance for Maxwell's neutral language:

> "If we can now explain the condition of a body with respect to the surrounding medium when it is said to be 'charged' with electricity, and account for the forces acting between electrified bodies, we shall have established a connexion between all the principal phenomena of electrical science" (Maxwell [1861], p.13).

So, bold blunt insertion of the molecular vortices theory into the electrostatic realm became possible due to the elasticity of the vortices that enabled the medium to maintain the elasticity waves. As a result,

> "the velocity of transverse undulations in our hypothetical medium, calculated from the electromagnetic experiments of M.M. Kohlrausch and W. Weber, agrees so exactly with the velocity of light calculated from the optical experiments of M. Fizeau, that we can scarcely avoid the inference that light consists of the same medium which is the cause of electric and magnetic phenomena" (Maxwell [1861], 1890, p. 22).

The introduction of displacement current was due to Maxwell's efforts to link the equations relating to electrical current with that of electrostatics. It demanded the Ampère law modification for the sake of a new term introduction; the term had to describe the elasticity of the vortices medium. The driving force due to introduction of displacement current owed its existence to Maxwell's strenuous efforts to unify all the main empirical laws belonging not only to electricity and magnetism but to optics as well.

As a result Maxwell obtained his immaculate system of equations along with the continuity equation describing the electrical particles that transform the rotations from one vortex to another. But one could not enunciate the final unification of optics and electromagnetism in 1861. It was possible to profess only commencement of their reconciliation, merely the beginning of 'grinding' of rather alien theoretical languages. And at last in 1864 Maxwell had proposed a modified version of his 1861 paper that avoided any special suppositions on the nature of molecular vortices. In his stupendous 1864 paper Maxwell derives his eminent equations from abstract dynamics of Lagrange.

Now he was able to derive the basic wave equation of electromagnetism without any screwy special assumptions about molecular vortices or forces between electrical particles. Although displacement still retained a prominent position in "*A Dynamical Theory of Electromagnetic Field*," its role differed from that it played in 1861 paper. It was no longer associated with changes in positions of rolling particles. Rather, Maxwell displayed it simply as the motion of electricity, that is, in terms of a quantity of charge crossing a designated area. Though, despite Maxwell's claim to provide deductions from (three) experimental facts, his account still required the postulation of a displacement current, something that could neither be verified by nor deduced from experiment (Morrison, 2000; Darrigol, 2001).

Finally, I suspect that an important role in 'grinding out' and 'cross-fertilization' of the British (field) and the Continental (action at a distance) research traditions was played by Maxwell's ethnic background. It enabled him to play the reverent role of intermediary in the communication processes. As a British citizen, Maxwell did obtain his scientific degrees at Cambridge. On the other hand, as a patriot of Scotland, he chose the colors of the Scottish national flag for his first color photograph. His critical comments on the British colonial policy in India are well-known too.

CHAPTER IV.

The Second Scientific Revolution Genesis: Planck's and Einstein's Piecemeal Matching of Pervasive Classical Research Traditions.

Looking back over the preceding, we may say briefly that the feature of the whole development of theoretical physics, up to the present, is the unification of its systems which has been obtained by a certain elimination of the antropomorphous elements, particularly the specific sense-perceptions.
— **Max Planck**

"*Fundamental crisis set in, the seriousness of which was suddenly recognized due to Max Planck's investigations into heat radiation (1900). The history of this event is all the more remarkable because, at least in its first phase, it was not in **any** way influenced by any surprising discoveries of an experimental nature*"
— **Albert Einstein**

Abstract.A generalized standpoint on the second scientific revolution naturally amalgamating quantum and relativistic 'mathematical projections of nature' is proffered. A subtle origin of the so-called 'old quantum theory' represented by Planck's 1897-1906 thorough papers is elicited. It is contended that, contrary to common belief, the quantum revolution of the early XXth century, at the origins of which was Planck, at least in its initial phase, was the result of not the revolutionary nature of its main hero, but rather of his conspicuous tendency to compromise. It was certainly Planck, who clearly realized profound cross-contradiction between 'the physics of material bodies and the physics of the ether' and outlined the first budding sketch of its withdrawal: the pervasive paradigms 'must be modified to remain compatible.' And it was Planck who took the first step in modifying the physics of the ether and contending that 'not only matter itself but also the effects radiated from matter' possess discontinuous properties, which can be characterized by a new natural constant: the elementary quantum of action. Einstein's necessary part consisted in that he took the next step in modifying the second component of the encounter – the physics of material bodies. Einstein's foolhardy 1905a light quanta hypothesis and subsequent distinctive 1905d special theory of relativity can be taken as mere milestones of the unwinding of Maxwellian electrodynamics and statistical thermodynamics reconcilement research program. The notorious conception of luminiferous ether turned out a substantial obstacle for Einstein's wayward statistical thermodynamics in which the pivotal lead was played by flagrant 1905a light quanta paper. Herewith Einstein was clearly aware that his enticing light quanta hypothesis was too audacious to be taken literally. Hence he laid out his 1905d version of 'electrodynamics of moving bodies' in a markedly Machian / Duhemian, phenomenological way. Though when related to creative momenta, his 1905 reconcilement *modus operandi* was initiated by a deft fusion of Machian empiricism and Planck's 'sophisticated realism' inspired by Kantian synthetic dualistic epistemology.

Key words: Boltzmann's statistical mechanics, Carnot and Clausius's thermodynamics, Maxwellian electrodynamics, quanta, relativity, Planck, Einstein, Duhem, Mach, Kant.

IV.1. Introduction. The Thrilling Ideal of Unity.

It is a platitude that Albert Einstein's multilateral scientific writings were strongly motivated by the Ideal of Unity of physical laws. Respectively, his scientific career, especially after the general theory of relativity (GTR) had been successfully contrived, presents an incessant search for unitary theories (see for instance van Dongen 2010). One should just recall Einstein's strenuous efforts to amalgamate gravity with other fields that lasted until his last April 1955 days. And it goes without saying that the incessant quest for dazzling 'Unity of Nature' is best represented precisely by *these* attempts of Einstein towards unitary theories during almost forty years than by his early bewitching writings.

However, in my view, Einstein's mature unification accounts and especially his superlative GTR arose out of his early works and first and foremost out of his 1905d successive attempts to contrive special theory of relativity (STR), as well as out of his audacious 1905a light quanta hypothesis.

Just to begin with, his startling 1905a light quanta paper apparently commences with highlighting "a *profound formal difference* between the theoretical conceptions physicists have formed about gases and other ponderable bodies and Maxwell's theory of electromagnetic processes in so-called empty space" (Einstein 1905a, p.86; my italics). The paper plainly aims at reconcilement of pervasive research traditions of classical physics.

Nay more, Einstein's 1905d STR paper also starts with unfolding a "*deep asymmetry*" (Einstein 1905d, p.140) in the common description of electromagnetic induction caused by a thorough gap between Newtonian mechanics and Maxwellian electrodynamics.

Onwards, as Einstein recalled later (Pais 1982), his perseverant 1907 efforts to set up the basic GTR tenet – the 'principle of equivalence' – were necessarily drawn upon his fresh impressions of contriving the STR.

Nevertheless, Einstein surely was not the first in ascertaining the glaring cross-contradiction between the pervasive research traditions of classical physics. According to Thomas S. Kuhn's 1978 thought-provoking monograph, at the end of the XIX-th century three staunch research programs were to *interact* (and one should specify – intertwine and interpenetrate) consequentially in Max Planck's creative work: thermodynamics, electromagnetic theory, and statistical mechanics. Before 1900, he has furnished substantial contributions to *all* three. But their weight ebbed away in his creativity from immaculate classical thermodynamics through Maxwell's new-fangled theory to Boltzmann's through and through controversial research programme. And thermodynamics was an infallible paragon of scientific theory for him.

Planck's corresponding advancements in it were celebrated and highly appreciated by the current scientific community before he first turned, not without substantial doubts and numerous hesitations, to odd realm of Maxwellian electrodynamics. Besides, one is not surprised by the eloquent fact that statistical mechanics intruded Planck's research much later and against stiff resistance. Accordingly, for Planck, the role of classical electrodynamics and statistical mechanics initially was purely instrumental. Maxwell's dainty equations and Boltzmann's sophisticated statistical technique proffered merely conceptual tools to solve the profound conundrum of irreversibility first and only then the inextricable technical puzzle of blackbody radiation.

Yet it was in no means accidental that it was Max Planck who ended up at the centre of the thorough clash between the classical physics basic paradigms. Primarily, as one of the first professional theoretical physicists, Planck was particularly sensitive, in Wien's wake, to intricate relations between different mellow physical research traditions. It was partly due to his 'sophisticated realism' Weltanschauung concerning the structure, dynamics, and progress of physical theories. The stern standpoint can be properly elicited via the comparison with Mach's confounded philosophical empiricism.

According to Planck, in the very beginning of science, the division of physics was grounded on the physiological peculiarities of man and possessed sheer anthropological character. Nevertheless, while originally the basic ideas of physics were taken from the specific sense perceptions of a common man, the latter are in modern times in large measure *excluded* from physical acoustics, optics, and theory of heat. Hence,

> "In short, we may say that the characteristic feature of the entire previous development of theoretical physics is a *definite elimination* from all physical ideas of the *anthropomorphic elements*" (Planck, [1910], 2015, p.4; my italics).

The peremptory result consisted in the attainment of 'unity and compactness in our system of theoretical physics.' On the contrary, due to the positivist blunt view, the pivotal problem of physics consists only in meticulous reading of the sense perceptions, following experience, to fixed laws. However, retorts Planck, though self-compatible, the positivist view has never contributed to any splendid advance in physics. It suffices to turn to the epoch-making scientific writings of Copernicus, Kepler, Newton, Huygens, Oersted, Faraday, Maxwell, et al. In a series of conspicuous lectures, eight in all, delivered with various amendments over a period ending about 1920 (and published by Dover) Planck alleged,

> "Looking back over the preceding, we may say briefly that the feature of the whole development of theoretical physics, up to the present, is the *unification* of its systems which has been obtained *by* a certain *elimination* of the anthropomorphic elements, particularly the specific sense-perceptions" (Planck, [1908], 1960, p.4; my italics).

For instance, the importance of Maxwellian electrodynamics consists in the hard fact that it integrated masterfully two branches of physics, which previously had to be treated as independent. That is why all the theorems which are valid for one branch, apply to the other as well.

> "Before the introduction of the electro-magnetic theory, physics was divided into three *separate* branches – mechanics, optics, and electro-dynamics, and *the unification of these is the ultimate and greatest aim of physical research…* and thus the number of independent branches has been reduced to two – the penultimate step towards the unification of the world picture. When and how the last step will be made, the linking up of mechanics and electrodynamics, cannot be said, and though many clever physicists are at present occupied with this question, the time does not yet seem ripe for the solution" (Planck [1908], 1960, p. 93; my italics;).

Correspondingly, Mach's celebrated epistemological doctrine, though extraordinary simple, clear and plain, represents, by and large, merely an understandable reaction against the proud expectations of the generations of physicists associated with universality of mechanical phenomena.

Eventually, Mach's pre-eminent and feasible principle of Economy of Thought

> "*is not the nature of science* and he [Mach] evades the most convenient of all scientific research – the finding of a *fixed* world picture independent of the variation of time and people…The *fixed unity* of the world picture is, however, as I have endeavoured to

show, the fixed goal which true science approaches through all its changes" (Planck, [1908], 1960, p.24; my italics).

That is why one should not be surprised to discover that *the red thread through all Planck's immaculate creative work was the problem of unification* of the mature 'paradigms' of the day.

> "Now, how far have we advanced today toward the unification of our system of physics? The numerous independent domains of the earlier physics now appear reduced to two: mechanics and electrodynamics, or, as one may say, *the physics of material bodies* and *the physics of the ether*. The former comprehends acoustics, phenomena in material bodies, and chemical phenomena; the latter, magnetism, optics, and radiant heat. But *is this division a fundamental one?* Will it prove final? *This is a question of great consequence for the future development of physics*" (Planck, [1910], 2015, p.4; my italics).

Note that Planck himself was quite optimistic concerning the future advancement of physics and vehemently tried to make a substantial impact of his own.

> "For myself, I believe it must be answered in the *negative,* and upon the following grounds: *mechanics and electrodynamics cannot be permanently sharply differentiated from each other.* Does the process of light emission, for example, belong to mechanics or electrodynamics? To which domain shall be assigned the laws of motion of electrons? At first glance, one may perhaps say: to electrodynamics, since with the electrons ponderable matter does not play any role. However, let one direct his attention to the motion of free electrons in metals. There he will find, in the study of the classical researches of H.A. Lorentz, for example, that the laws obeyed by the electrons belong to the kinetic theory of gases than to electrodynamics. In general, it appears to me that *the original differences between processes in the ether and processes in material bodies are to be considered as disappearing. Electrodynamics and mechanics are not so*

remarkably far apart, as is considered to be the case by many people, who already speak of a conflict between the mechanical and electrodynamic views of the world" (Planck, [1910], 2015, p.4; my italics).

Note that, on Planck's experienced and sophisticated view, Newtonian mechanics and Maxwellian electrodynamics are not *incommensurable*. Mechanics is grounded on the ideas of space and time, but just the same can be said about the classical electrodynamics. Hence

> "In fact, there are many indications pointing toward the *ultimate amalgamation* of these two subjects, the domains of which already overlap in some measure" (Planck, [1910], 2015, p.4; my italics).

Consequently, Planck considered himself as a man who strives to provide the situation when

> 'the gulf between ether and matter will be once bridged' answering the exciting questions that "will characterize the whole nature of the further development of our science" (Planck, [1910], 2015, p.4).

In light of the above, it is not surprising that

> "although Planck was intimately associated with the two major innovations in physics in the 20th century, quantum physics and relativistic physics, he never intended to participate in a revolution in physics" (Goldberg 1976).

Nevertheless, gritting his teeth, he was forced to do it

Hence *the overall aim of the present chapter is to take the next available step and to unfold the abiding influence of the abovementioned reconcilement on Planck's 1897-1900 and Einstein's 1902-1905 startling writings and especially on STR intricate genesis and advancement.*

Forsooth, commenting on Thomas Kuhn's distinctive view of paradigm change, Joseph Agassi asserted that

> "First and foremost, the very idea that scientific research rests on conversion, that conversion of import for research may be vague

> and quite general, yet later become more specific. We have so many examples of researchers who tried their hands in diverse directions, who developed ideas which were quite contrary to their deepest convictions, who developed ideas of their opponents in the spirit of self-criticism, in the spirit of criticism, or both. It is time to stop writing the history of scientific research while using religious metaphors" (Agassi 1983, p.379).

Respectively, the important part of this chapter deals with the circle of matching problems that irrepressibly brought Einstein to the intricate domain of 'electrodynamics of moving bodies.' Correspondingly, the subsequent part aims at answering the question: what was the runaway train of thought that engendered Einstein to contrive light quanta and STR on the strength of Planck's promising advancements. It is contended that the former and the latter turn out to be mere milestones of implementation of Maxwellian electrodynamics, statistical mechanics and thermodynamics reconciliation program that originated in Planck's pioneering writings. The pivotal role in the synthetic program was played by Einstein's 1905a light quanta hypothesis, since it was certainly the ether conception that put insuperable difficulties in realization of Einstein's statistical-thermodynamics design. Einstein was fully aware that his quaint light quanta hypothesis was audacious and that he needed to be especially cautious. Hence he carefully posited his 'electrodynamics of moving bodies' in a markedly Duhem-Mach phenomenological way. In particular, Einstein hung out his basic relativity postulate as originating from the Stevin-Mach 'instinctive knowledge.' (Namely this peculiarity may explain the difficulties in the acceptance of the STR abroad, for instance in Britain – see Goldberg 1970).

Nevertheless, when related to creative momenta, his 1905 reconcilement *modus operandi* was initiated by a deft fusion of Machian empiricism and Planck's 'sophisticated realism' typified in Kantian synthetic dualistic epistemology.

IV.2. Einstein as a Young Man and the Lumineforous Ether.

In good old Germany of the second half of the XIXth century Maxwell's strenuous efforts to concord the research programs of Young-Fresnél, Faraday and Ampére-Weber (see the preceding chapter; see also Nugayev 2015) were reinforced by Berlin university eminent physicist and physician, as well as keen philosopher of science Hermann Ferdinand Helmholtz. In Helmholtz's conspicuous paradigm charges and currents were punctiliously taken as the true sources of electrical and magnetic fields. Eventually it led to Hendrik Lorentz's 1892-1900 dualistic worldview grounded on the sober conjunction of the field equations and the equations of motion. Not by chance Lorentz's masterful theory was a prolific amalgamation of Maxwell's field theory and Wilhelm Weber's deft particle theory of pre-Maxwellian electrodynamics.

And it was a brave young scholar Albert Einstein who dared to pick the problem up after such boffins. Exactly this an ETH (Eidgenossiche Technische Hochschule) ingenious student openly admitted to his fiancée in 10 August 1899 'Paradies' hotel letter (Doc. № 52 of Einstein 1987, p.131).

From the very start of his dazzling scientific career Einstein had repeatedly expressed sober doubts on the role of 'des Namens Aether' in classical electrodynamics. Yet his reasonable skepticism was selectively directed at Hertz's concept of the ether as *a medium with a certain state of motion*, but not at the renounced ether concept *itself*. Einstein pronouncedly attributed major significance to the concept of 'elektrische Massen' and took electric currents as *real* motions of such charges in empty space, not as the 'Verschwinden elektrische Polarisation in der Zeit' (disappearing electrical polarization in time). At the beginning of Einstein's scientific career his *Weltanschauung* was in the most common way provoked by the fine lectures on electricity of his ETH physics teacher prof. H.F. Weber, as it is unfolded by Einstein's lecture notes (see Doc. № 37 and the related comments in Einstein 1987, pp. 223-225).

The 'substantive' concept of electricity was worked out by Wilhelm Weber and was widely used by many German-speaking physicists, including H.F. Weber. Therein, at least *initially Einstein's common views on electrical masses moving in the immobile ether were quite similar to H.A. Lorentz's eminent dualistic paradigm.* Withal, it was H.F. Weber who stimulated Einstein's keen interest, apart from the atomistic constitution of matter, to the relationship between the thermal and the electric properties of matter. Not by chance Albert's diploma thesis was dedicated to heat conduction (Renn 1993).

As early as 1899 Einstein became clearly aware of the conflict between the assumption of a stationary ether and the relativity principle of classical mechanics (Einstein's letter to Maric, 10 August 1899, Einstein 1989, Doc.52; see also Renn 1993). At the end of 1899 he had punctiliously studied a paper by Wilhelm Wien (Wien 1898) from which he learned about lack of any experimental evidence for the ether wind which should be an inexorable consequence of the assumption of a stationary ether.

Respectively, Einstein concluded the abovementioned letter recapitulating that 'Strahlungversuche' (radiation experiments) was needed for choosing between the two standpoints he outlined. And his next, 10 September 1899 'Paradise hotel' letter to Marić exhibited a sound idea for experimentally ascertaining the influence of motion relative to the ether on the propagation of light in transparent bodies.

Regrettably, Einstein's professor of physics exposed no enthusiasm for his work, and so Albert made no further mention in his papers and correspondence of his setting force the electrodynamics of moving bodies *for almost two years*.

However, 'die prinzipielle Trennung von Lichtaether und Materie' (the fundamental separation of light ether and matter), 'Definition absoluter Ruhe' (definition of absolute rest), and so on of course were among the medley topics he strenuously discussed with his close friend Michele Besso (see Einstein's 4 April 1901 letter to Marić). In March 1901 Albert confidentially informed Mileva that he looked forward to the perfection of "unsere *Arbeit uber die Relativbewegung*" (our work on relative motion). In September 1901 he proudly informed his another close friend and former classmate Marcel Grossman on working up a simpler method for the investigation of the motion of matter relative to ether, based 'auf gewonlichen Interferenzversuchen' (on ordinary interference experiments). By December 1901 he was 'arbeite eifrig' (working hard) on "*die Elektrodynamik bewegter Körper*," that promised to become "eine kapitale Abhandlung" (Einstein's 17 December 1901 letter to Marić). A sad calculation mistake had earlier led him to doubt the correctness of his startling 'Ideen über die Relativbewegung,' but he now believed in these ideas 'even more.'

He elicited the motley stuff to prof. Kleiner and the latter even

> "thought that the experimental method proposed by me is the simplest and most appropriate and conceivable. I was very pleased with the success. I shall certainly write the paper in the coming weeks" (Einstein's letter to Marić, 19 December 1901, p. 189).

Despite of prof. Kleiner's sage encouragement and Einstein's youthful enthusiasm, *no publication on this subject ensued for over three years – till 21 June 1905.* - Why? – After all, Einstein was really engaged in working on 'eine kapitale Abhandlung' (capital treatise) on the electrodynamics of moving bodies at the end of 1901. Then he had stopped doing it and retraced to the manuscript only in 1905. What happened at that time, and *why* had Einstein, being initially a moderate supporter of the ether, became its committed enemy?

To come to a judicious explanation one has first to recall Einstein's derogative evaluation of his youth writings – 'my worthless beginner papers' (Einstein / Marić 1992). The evidence at hand highlights that the planned precocious "*kapitale Abhandlung*" was a 'far cry' from the 1905d conspicuous STR paper. On the contrary, now one definitely knows (Rynasiewicz 2000) that Einstein had arrived at the body of startling results of his 1905d relativity paper, in a 'sudden burst of creativity' and *only after he had perfected his first three works in the spring of 1905.* The key insight – the discovery of the relativity of simultaneity – occurred to Einstein *only in late May 1905* and only *after* the completion of the 1905c Brownian motion paper. For example, when asked by the guileless biographer Carl Seelig, Einstein elucidated:

> "Between the conception of the idea of the special theory of relativity and the completion of the corresponding published paper

there passed five or six weeks" (Seelig 1960, p.114).

IV.3. Max Planck's trailblazing impact.

To provide a weighty answer one has to dip first into the STR paper itself (Einstein 1905d). It famously commences with unfolding a '*deep asymmetry*' in theoretical reproduction of the phenomenon of electromagnetic induction. Experience holds that the induction current engendered in the conductor by the motion of the magnet depends only on their *relative* motion. On the other hand, the Maxwell-Lorentz theory contrives *two substantially* different accounts of the effect that mysteriously lead to *one and the same* quantitative result.

Yet note that *Albert Einstein was by no means the first* to discern the substantial asymmetries in theoretical representation of the induction phenomenon (Darrigol 2001, p.377).

Hence the thrilling question is not how Einstein became aware of the asymmetries, but *what made them so intolerable to him.* In my humble opinion, the key to answer the above question lies in *other* works of Albert Einstein and first of all in his medley 1905 papers. It is well-known that Einstein published *nothing* on the topic of optics and electrodynamics of moving bodies prior to 1905. More importantly, it was Albert Einstein himself who had just revealed *another asymmetry – and of more substantial kind* – in other startling 1905a paper "*On an heuristic point of view concerning the processes of emission and transformation of light.*" The paper was published in the same lead magazine "*Annalen der Physik*" but three months ***before*** the STR paper. Behold the outset of his 1905a ground-breaking work:

> "There exists a *profound formal difference* between the theoretical conceptions physicists have formed about gases and other ponderable bodies and Maxwell's theory of electromagnetic processes in so-called empty space" (Einstein 1905a, p.86; my italics).

Nevertheless, to comprehend how Einstein arrived at profound 'cross-contradiction' (Nugayev 1999) between mechanics and electrodynamics one has to precipitate himself to Max Planck's pioneer writings first.

At the turn of the XIX-th century, three thorough 'classical' paradigms had interacted strongly in Max Planck's oeuvre: thermodynamics, electromagnetic theory and statistical mechanics (Kuhn 1978). And initially the role of electrodynamics and statistical mechanics was collateral for Planck. Maxwell's startling equations and Boltzmann's subtle technique provided merely conceptual instruments to solve the fundamental problem of irreversibility first; and only then to grapple with the blackbody radiation.

Note that even proper classical theory of black-body radiation before Planck's vehement efforts did not exist at all. Planck's sedulous writings represented practically the first effort to contrive a true *theory* of black-body radiation.

Viz., they constituted a bold attempt to rederive the radiation density formula $u(\nu, T) = \acute{\alpha}\, \nu^3\, [\exp(b\, \nu/T) - 1]$ (*) directly from the fundamental *principles* of thermodynamics and electrodynamics as opposed to phenomenological efforts to 'guess' the correct expression through unsteady experimental data. All the pre-Planck blackbody radiation results had in his eyes, even in the best (*) case, "only a very limited value - only as happily guessed interpolation formulas" (einer glucklich erratenen Interpolation formel – see Dmitriev 2019).

> "Therefore, - Planck recalled, - from the day it [*] was found, I was busy with the task of establishing its true physical meaning, and this question led me to consider the connection between entropy and probability, i.e. to the Boltzmann way of thinking" (Planck 2001, p.29).In 1931 R. Wood letter Planck admitted that "theoretical explanation of the formula [*} had to be < > found at any cost, and no price would be too high" (eine theoretische Deutung < > um jeden Preis gefunden warden [musste], und wäre er noch so hoch) (quoted from Dmitriev 2019, p.7).

Accordingly, Planck met with many obstacles on this thorny path, perseverantly deriving partial theoretical laws, comparing them with (not always reliable) experimental results, correcting the laws and checking the consequences, and so on.

And Planck was a theoretician 'par excellence' with thermodynamics taken as a paragon of a mature scientific theory. Already in Planck's times thermodynamics has grown to an established and respectable research tradition with a multitude of valuable practical applications. Yet when he had ascertained that thermodynamics was insufficient, he started to apply Maxwellian electrodynamics – rather young and controversial at his times, whose proper empirical verification and technical application had just commenced. Nevertheless, Planck, in his quest for comprehension of irreversibility, had to modify his theoretical scheme at the expense of inserting subtle abstract objects into it. And thus he had constructed resistanceless ideal vibrating resonators– tiny oscillating currents governed by Maxwell's equations. Note that Planck's ideal constructs had nothing to do with real experimental devices. They did not carefully mimic actual molecules, atoms or even electrons. Rather, they were peculiar theoretical contraptions conjured up to set up the theoretical model establishing thermodynamical equilibrium between radiation and matter.

In 1897 Planck had gingerly published the first paper of a five-part series entitled "*On Irreversible Radiation Processes*" (Planck, 1897a). In all the five writings (Planck, 1897, a, b, c; 1898, 1899), as well as in the basic article, which recapitulated their results for the "*Annalen der Physik*" (Planck, 1900b), he had scrutinized the attributes of an ideal model consisting of a system of resonators intensively interacting with an electromagnetic field.

However, after he had read his first paper to the Academy, a scathing criticism by Ludwig Boltzmann himself was proffered to the same audience. Boltzmann contended that, though Planck's complex formulas for resonator absorption and emission were immaculate, the reductionist 'irreversibility' program for which they were designed should pronouncedly fail. Indeed, both Maxwell's equations and the boundary conditions on their solution are implacably invariant under time reversal.

When the application of Maxwell's equations turned out to be insufficient, Planck had to implement statistics – quite unwillingly, gradually, and under the growing pressure of hard experimental results now. The following delightful story is especially pertinent here (Dugas, 1959; De Broglie, 1962, p.139).

As a representative of German school, Max Planck stubbornly intended to broaden the domain of validity of statistical thermodynamics. So he advanced (in the wake of Wilhelm Wien, 1893) thermodynamics of electromagnetic radiation and strived to introduce the entropy of radiation by analogy with its energy. Being a partisan of eminent Ludwig Boltzmann, Planck informed the luminary, a father of statistical mechanics, about the research, and even dared to present one of his papers to patriarch's strict judgement. Alas, Boltzmann harshly retorted that Planck would never be able to create a viable statistical thermodynamics of electromagnetic field *without inserting previously unknown element of discontinuity into processes of radiation.*

Already in 1872, 28 years *before* Planck's ground-breaking 1900b paper, Boltzmann in his "*Further investigations of thermal equilibrium between gas molecules*" pioneered in applying the notions of discrete energy to the exchange processes.

Contemplating the second law of thermodynamics from a statistical point of view, Boltzmann arrived at the notion of 'energy atoms' transpiring themselves in the intricate processes of molecular interactions. The queer notion of the finite energy amounts that can be freely exchanged for colliding molecules led Boltzmann to complicated calculations of the number of collisions with the help of subtle combinatorials technique. Nay, due to the prevailing Weltanschauung, Boltzmann posited the notions of energy quanta just as an auxiliary mathematical trick (similar to complex numbers).

Though note that the father of statistical mechanics hold more profound views on the subject (Bordoni 2013). In a lecture red in 1904 in St. Louis, at the Congress of Arts and Science, he maintained that "the battle which the theories [mechanics vs. electrodynamics] have to fight is, however, an infinitely wearisome one." This is so much so since theoretical physics dealt with "certain disputed questions which existed from the beginning" and which "will leave as long as the science." A vivid example is the choice between discrete and continuous in the representation of matter.

Consequently, physical theories should not be taken as 'incontrovertibly established truths' since they are grounded on the hypotheses which 'require and are capable of continuous development' (Boltzmann 1905, pp.592-5).

Thus, the 'Boltzmann heritage' turned out to be especially controversial and debatable. At first Planck tried to get through by using an electromagnetic analogue of Boltzmann's H-theorem. And when it definitely failed, he had to appeal to the probability calculus and intricate combinatorial definition of entropy which he wholeheartedly hated (Kuhn, 1978).

By the beginning of 1900, only one facet of Boltzmann's treatment of irreversibility was still absent from Planck's successful research program, the application of combinatorials. And by the end of the year Planck had embraced it, too. Though what forced him to do so was no longer the intricate problem of irreversibility. On the contrary, it was the passionate search for a radiation law that could like a house on fire pass the test of new, more sophisticated tests.

The apparent weakness of the derivation of the Wien distribution law, that Planck submitted to the '*Annalen der Physik*' in November 1899, was the lack of a uniqueness proof for the function he had assigned an oscillator entropy. Yet pointing on their reliable measurements of the frequency distribution of radiation from a new piece of experimental apparatus, the first laboratory black cavity, Otto Lummer and Ernst Pringsheim arrived at a new formula for the blackbody radiation that deviated from Wien's law significantly. But Planck had luckily eliminated the difficulty in a paper submitted for publication in February 1900, in which he asserted to have *derived,* rather than defined, oscillator entropy for the first time, thus supporting Wien's law again.

However, the experimental results turned out to be persistent. In a paper, reported to the Physics Section of the Naturforscherversammlung on the 18 September 1900, Lummer and Pringsheim maintained that the Wien-Planck distribution law again did not represent their perseverant measurements on black radiation in the low frequency region. In the range, achieved with the aid of recently advanced techniques, the discrepancies between experiment and theory were near 50% and blatantly could not be due to simple experimental errors.

The evidence was rather convincing, but Planck was well prepared to meet it now. In an important paper presented to the Physical Society on 19 October 1900 -he shrewdly referred to the proof of the Wien law he had submitted to the '*Annalen der Physik*' in March, and properly underscored its shortcoming. The entropy of n oscillators, he asserted, must depend not simply on their total energy, as was surmised, but on the energy U of a single oscillator.

The expression $\delta^2 S/\delta^2 U$ = - α/U derived earlier is too blunt and should be necessarily exchanged for a more sophisticated one. Planck had proffered an expression that

> "is the simplest by far of all the expressions which yield S as a logarithmic function of U (a condition which probability theory suggests) and which besides coincides with the Wien law for small values of U."

If the equation for S is taken as the first term (-U/α) in a power series expansion of the expression $(\delta^2 S/\delta^2 U),^{-1}$ his new form follows immediately by the addition of a term proportional to U^2.

With the renewed expression $\delta^2 S/\delta^2 U$ = - α/U(U+β), two integrations, and the standard condition $\delta S/\delta U$ = 1/T an application of the displacement law renders a new distribution law U = bν/ exp (αν/T) – 1. And this radiation formula, Planck proudly opined,

> "so far as I can see by quick inspection, represents the hitherto published observational data just as *satisfactorily* as the best previously proposed distribution function…I therefore feel justified in directing attention to this new formula, which, from the standpoint of electromagnetic radiation theory, I take to be the *simplest* excepting Wien's."

New refined measurements by O. Lummer and E. Pringsheim quickly exhibited the last equation to be superior to all the other known distribution laws.

> "The following morning I received a visit from Rubens. He told me that after the meeting, the same night, he had compared my formula with the results of his measurements and had found good

agreement in every point. Lummer and Pringsheim too" (Planck 1948, p.26).

Yet Planck had to envisage the route to the formula that was less ad hoc. As Thomas Kuhn shrewdly pointed out, this has commonly been comprehended as Planck's ponderous *conversion* from phenomenological thermodynamics to a statistical one. But the important breakthrough had occurred at least a year, and more probably three years, *before*. When Planck referred to stupendous 'relation between entropy and probability,' he did not have in mind the statistical approach in general but only, as his words openly suggested, *Boltzmann's startling combinatorial definition of entropy*.

However, Planck's primordial derivation of the Wien law contained a conspicuous internal contradiction. The n resonators he contemplated were required to be independent, but his argument depended on supposing that their total energy Un was distributed *equally* among them. A more perfect argument would uncover the various ways in which that energy might be allocated between resonators just as Boltzmann, in his pioneer combinatorial studies, had divided the total energy of a gas among its molecules. Expression for Planck's distribution law $U = b\nu/ \exp(\alpha\nu/T) - 1$ can be dexterously manipulated to yield $1/T$ as a function of U and ν, and $1/T$ is just $\delta S/\delta U$. After integration one inexorably gets

$$S = (b/a) \log \{(1 + U/b\nu)^{1+U/b\nu} / (U/b\nu)^{U/b\nu}\} + \text{const.}$$

Planck should have been amazed by its apparent resemblance to Boltzmann's epoch-making expression for the logarithmic relation between entropy and probability.

Alack, the equation applies only to a single resonator with energy U in equilibrium with a radiation field and is not suitable for proper interpretation in probabilistic terms. Thus, let one imagine N independent resonators of frequency ν set up in equilibrium with their radiation field. Their total entropy should be equal to NS, while their total energy should be equal to NU. If odd combinatorials are to be engaged, the total energy must be subdivided into P elements of size ε, so that $P\varepsilon = NU$. Multiplying Planck's equation for S by N and substituting $P\varepsilon/N$ for U yields

$$SN = (b/a) \log \{(N + P\varepsilon/b\nu)^{N+P\varepsilon/b\nu} / N^N (P\varepsilon/b\nu)^{P\varepsilon/b\nu}\} + \text{const}$$

To obtain an expression involving only integers, the size of the energy element ε must be set equal to bν. The quantity in { } then reduces, for large N and P, to (N+P-1)!/ (N-1)!P! But that formulae is the common expression for the number of ways in which P indistinguishable elements can be distributed over N boxes. Ergo, the first stages of Planck's application of Boltzmann's relation between entropy and probability were completed in this way. But the intricate problem still remained. The combinatorial expression detected by working backwards from Planck's distribution law decidedly differed from the one Boltzmann had obtained in deriving the equilibrium distribution of gas molecules.

Hence Planck had to elicit that it is proportional to the probability appropriate to equilibrium radiation. His ultimate problem was to deduce the expression for the entropy of a particular distribution of the total energy E over N resonators. Then he had to find its maximum with respect to variation of the distribution of the total energy over frequency.

To find the entropy of an arbitrary distribution Planck was forced to employ combinatorials and hence he diligently followed Boltzmann in subdividing the energy continuum into the elements of finite size (Planck, 1900b; translated by Ter Haar, 1967, p.83).

Then Planck defines a 'complexion' (an expression, he punctuates, 'applied by Boltzmann for a similar concept') as a particular specification of the set of numbers, which fixes the number of elements ε attributed to the various resonators in the set of N. The total number of possible complexions is R. To determine the equilibrium distribution one is compeled to maximise R or log R by varying the energies at the various frequencies. Straightforward manipulations exhibit that the entropy (log R) will have a maximum if $U_\nu = h\nu/\exp(h\nu/kT) - 1$. The corresponding expression for the distribution of the field is $u_\nu = (8\pi\nu^2 / c^3)\, U_\nu$.

It should be stressed that both in his original derivation papers and far more clearly, in his superlative «*Lectures*» (1906), Planck's radiation theory is *incompatible* with the *actual* quantization of the resonator energy.

His theory timorously requires fixing the size of the small intervals into which the energy continuum is subdivided merely for purposes of combinatorial computation. In Planck's odd theory resonator emission and absorption are still traditionally governed by Maxwell's equations. Indeed, Planck did repeatedly employ expressions like UN = Phν.

But UN is the total energy of N resonators. Restricting it to integral multiples of hν does not *necessarily* impose any similar restriction on the energy of an *individual* resonator. It may still vary continuously.

So, though Planck incessantly appealed to the reliable experimental results, the role of the experiment in creating his theory should not be overestimated. Black-body experiments really played the role of the important factor. But they only forced Planck to implement statistics in the growing rates. In the lack of experimental data Planck would not apply Boltzmann's combinatorials in full rate he did not like them wholeheartedly. The implementation of combinatorials was necessarily connected with inserting such distinctive and dubious hypotheses that Planck, an admirer of classical thermodynamics, vehemently tried to evade.

Thus one cannot draw a common conclusion that Planck's epoch-making distribution law was a *direct* generalization of experimental results. On the contrary, Planck's plod to it was 'from top to bottom.' Surely, he had to take the experimental evidence into account. Yet it forced him not to 'deduce from phenomena' but to apply more theory instead. As a genuine theoretician, Planck was fairly sensitive to the gamut of the problem he tried to solve and to the necessity of taking it in the subtle 'intertheoretic context.'

He clearly comprehended an origin of the problem lying in the pervasive contradictions between mechanics, statistics, electrodynamics and thermodynamics (Planck, 1910, p.616; my italics. See especially Planck's immaculate 1906 "*Lectures on Heat Radiation*," 1906, p.105).

Planck understood that the introduction of $\varepsilon = h\nu$ sprung out of the gaping abyss between Boltzmann's statistical mechanics and Maxwell's classical electrodynamics. Eventually the ultimate elimination of the cross-contradiction resulted in the contrivance of quantum electrodynamics, i.e. in invention of quantum theory of radiation that considered electromagnetic particles as Boltzmann's molecules that can gain energy under collisions with usual molecules and resonators. Though an important reservation should be made typified in the following Eugene Wigner's quotation:

> "The statement, that there are no conceptual conflicts between quantum mechanics and the special theory [of relativity], should not mean that the mathematical formulations of the two theories *naturally mesh.* This is not the case, and it required the very ingenious work of Tomonaga, Schwinger, Feynman, and Dyson to adjust quantum mechanics to the postulates of the special theory [of relativity] and this was so far successful only on the working level. What is meant is, rather, that the concepts which are used in quantum mechanics, measurements of the position, momenta, and the like, are the same concepts in terms of which the special relativistic postulate is formulated.
>
> Hence it is at least possible to formulate the requirement of special relativistic invariance for quantum theories and to ascertain whether these requirements are met. The fact that the answer is more nearly no than yes, that quantum mechanics has not yet been fully adjusted to the postulates of the special theory [of relativity], is perhaps irritating" (Wigner 1957, p.109).

Hence, it is no coincidence that the next felicitous scholar to notice the profound cross-contradiction abovementioned appeared to be Albert Einstein. The next hallmark he famously inherited from Planck was his obsession with mature theory unification. Existence of the Planck-Einstein link is supported by the following Michele Besso's letter to Albert Einstein dated January 17, 1928:

> "As far as I'm concerned, I was your audience in the years 1904 and 1905; if, in the drafting of your publications on the quantum problem, I deprived you of some of your glory, in exchange I provided you with a *friend in Planck*" (Einstein and Besso, 1972, p.238).

IV.4. *What was the bolt upright train of thought that brought Einstein to light quanta and then to special relativity?*

In the first part of the 1905a paper Einstein exhibited that the *joint* application of mechanical and electrodynamic 'theoretical pictures' for elucidating the black-body radiation leads not only to the apparent contradictions with experiments (his paper did not even cite Lummer & Pringsheim or Rubens & Curlbaum veracious results), but to the staggering *paradox* that cannot be wriggled out by common means. To elucidate it, Einstein contrives the gedankenexperiment with the both theories. He contemplates an imaginary cavity containing free electromagnetic field, gas molecules and Planck's ideal resonators and arrives at a conclusion that the joint application of mechanics and electrodynamics *unavoidably* leads to Rayleigh-Jeans law for energy density of the black-body radiation.

> "This relation which we found as the condition for dynamic equilibrium does not only lack agreement with experiment, but it also shows that in our picture there can be no question of a definite distribution of energy between aether and matter,"
>
> since
>
> "the greater we choose the range of frequencies of resonators, the greater becomes the radiation energy in space and in the limit we get $\int_0^\infty \rho_\nu \, d\nu = (R/N)\,(8\pi/L^3)\,T \int_0^\infty \nu^2 \rho_\nu \, d\nu = \infty$."

Although it is commonly held that in the 1905a light quanta paper Einstein was obsessed with an explanation of the photoelectric effect, the study of the masterpiece unfolds that this was not the case. Measurements of the effect at that time were not sufficiently accurate to point definitely to a violation of classical behavior (Ter Haar 1967). Hence Einstein was preoccupied not so much by the evidence relating to photoeffect and accordingly appealed to fluorescence, photoelectricity and photoionization data only as to *indirect* evidence in favor of his frail model.

On the opposite, Einstein had mostly delved into cogitation of the profound *contradiction* between statistical mechanics and Maxwellian electrodynamics, as well as looking for the efficacious ways out of it.

So, *what was a lofty reason of Einstein's sincere interest* to the contradictions between the mature classical physical theories? - To find out a judicious answer one has to grapple with Einstein's 1946 '*Autobiographical Notes*' once more. Namely this source exhibits the most extensive and systematic recital of evolution of Einstein's views (thereunto made by Einstein himself):

> "in my younger years, however, Mach's epistemological position also influenced me greatly…" (Einstein 1949a, p. 21).

Note that Mach's epoch-making "*Science of Mechanics*" is teeming with stubborn attacks against classical mechanics' asset in physics. In Mach's wake, Einstein could therefore freely, easy and fun juxtapose Newtonian mechanics, Maxwellian electrodynamics and statistical thermodynamics without reducing one to the others. Yet the pivotal element of Machian epistemology that persistently accompanied Einstein all his life was Mach's hardy *Principle of Economy of Thought*: "Physics is Experience Arranged in Economical Order" (Mach 1897/1984). Thereafter, in punctilious review of the STR genesis, published in "*Science*" in 1940, Einstein admits that

> "the theory of relativity arose out of efforts to improve, with reference to *logical economy*, the foundation of physics as it existed at the turn of the century" (Einstein 1940/1954, p. 329; my italics; see also Einstein 1933/1954, p.277; Einstein 1944/1954, p. 23).

And a proper explanation of Einstein's clamorous reasons for writing his 1905a paper and its connections with the other 1905 ones can be discovered in his "*Autobiographical Notes*" again. Due to Einstein, the first stage of 'the revolution [that had] begun by the introduction of the field' (Einstein 1949a, p. 37) consisted in the invention and deployment of the Maxwellian electrodynamics. All the pre-Maxwellian deft accounts of physical interactions (the conspicuous theories of Newton, Ampére,

Weber, Riemann et al.) were theories of interactions between several *material points*.

To be precise, owing to Faraday and Maxwell, the *Electromagnetic Field* was inserted into the texture of the XIX-th century physics as an element of physical reality tantamount to the *Material Point*. The problem situation was characterized by long-drawn

> "dualism which lies in the fact that the material point in Newton's sense and the field as continuum are used as elementary concepts *side by side*. Kinetic energy and field-energy appear as *essentially different* things" (ibid, p.37; my italics).

It is crucial that, as an incorrigible consequence of the dualism,

> a "*fundamental crisis* set in, the seriousness of which was suddenly recognized due to Max Planck's investigations into heat radiation (1900). The history of this event is all the more remarkable because, at least in its first phase, *it was not in any way influenced by any surprising discoveries of an experimental nature*" (ibid, p.37; my italics).

Max Planck's wondrous way of reasoning ($\varepsilon = h\nu$) defiantly contradicted the mechanical and electrodynamical basis upon which his derivation depended. Note that according to Einstein himself

> "My own interest in those years was less concerned with the detailed consequences of Planck's results, however important these might be. My major interest was: What general conclusions can be drawn from the radiation formula … concerning the structure of radiation *and even more concerning the electromagnetic foundations of physics?*" (Einstein 1949a, p.47; my italics).

Hence Einstein's close attention in the 1905a paper to the subject of theory of quanta was spurred by its reconciliation possibilities, for its breathtaking prospects to arrive at a successful *fusion* of Maxwellian electrodynamics and Boltzmann's statistical thermodynamics. Herewith Einstein was fully aware that his quantum hypothesis was too audacious to be taken literally. Next we will exhibit that he definitely did not maintain that "light quanta of energy ε = hν really exist." His formulation was more cautious: it would be *possible* to expose monochromatic radiation in certain limits *as if* it were composed of mutually independent energy quanta. Nevertheless, already in 1907 Planck contended:

> "Does the absolute vacuum (the free ether) possess any atomic properties? According to your remark…that the electromagnetic state in a [finite] region of space is determined by a finite number of magnitudes, you appear to respond affirmatively to this question, while I would, according to my present views, disagree. Because I search for the meaning of the elementary quantum of action (light quantum) not in the vacuum but at the sites of absorption and emission, and I assume that the processes in the vacuum are exactly described by Maxwell's equations" (Planck to Einstein, 6 July 1907, The Albert Einstein Archives, The Hebrew University of Jerusalem, 19246. Quoted from Renn 1993, p. 318).

Probably the difference between Planck's and Einstein's standpoints can be understood as follows. To comprehend Einstein's true attitude to light quanta surmise in the context of its fusion capacities one should start from the following quotation from Jeroen van Dongen's bona fide book "*Einstein's Unification*":

> "on a number of occasions Einstein actually expressed himself quite appreciative of Kant's ideas, and some aspects of Einstein's thought did rather resemble the Kantian philosophy. Both for instance emphasized the virtue of striving for *unity in science*" (van Dongen 2010, p.49; my italics).

And, in my view, *for the milestones of Einstein's 1905 oeuvre Kant's influence should be taken as crucial.*

To begin with, the very possibility of Kant's substantial influence on 'Einstein as a young man' is apparent: Kantian philosophy was undoubtedly prevalent among the educated classes in Germany in the late XIX-th century. And the extent to which it was rife in German high school was 'overwhelming' (Beller 2000, p.84). Respectively, neo-Kantianism

('Neukantianismus' with its succinct motto: "Zurück to Kant!") was the common philosophical movement in German high school from the 1870's until the First World War. Needless to say that Einstein first read Kant at the age of thirteen and again at the age of sixteen (Howard 1994, p.49).

Thereafter, being an Eidgenössiche Technische Hochshule (ETH) student in Zurich, he had an excellent opportunity to continue his Studium von Kant. Kant's Nachlaß was punctiliously disclosed at the lectures of Dr. August Stadler (Einstein 1987, pp. 45-50), a prominent neo-Kantian of Marburg school and withal Hermann Cohen's first doctorate.

What on earth could bewitch 1905 Einstein in murky waters of Kantian epistemology? According to Kant, in the long run it is our freedom from the world that makes science possible. The sensible world eventually conforms to basic laws because the human mind artfully *constructs* it according to certain laws.

It is constructivist foundation for scientific knowledge that provides that a priori knowledge of 'sache an sich' (thing in themselve) is impossible.

Yet in the *Appendix to the "Dialectic" of the first Critique* Kant strived to rehabilitate the bored ideas of common metaphysics by asserting that the ideas of reason play a significant role in the conduct of natural science only if they are comprehended *regulatively*. Viz., if they are taken to represent not metaphysical beings or entities, but rather ultimate *goals and directions* of scientific enquiry.

In the "*Critique of Pure Reason*" Kant implacably divided the human intellect into the independent faculties of (1) sensibility, (2) understanding, and (3) reason. Respectively, the principles governing these faculties belong either to (I) *constitutive* principles, or to (II) *regulative* ones.

I. The constitutive principles were taken as strict rules for the construction of the phenomenal world (e.g. Newton's laws, Maxwell's laws, etc.). Thus, pivotal distinction between the constitutive and regulative principles appears to be drawn in Kant's chef-d-euvre along the following lines. The constitutive principles are those that govern the function of understanding and are necessary conditions of experience, whereas regulative principles – the 'Ideas of Reason' –

dominate the function of reason and are not instantiated in experience in the same way.

By the way, it was Hans Reichenbach who started to contemplate constitutive principles as *synthetic a priori* ones. He maintained that to retain a role for such principles they should be *relativized* and comprehended as a priori only from the perspective of a particular mature theory like Newtonian mechanics or Maxwellian electrodynamics.

Though constitutive principles are the necessary conditions for the possibility of the experience, they do not necessitate our own experience of the world. Consequently these principles define the space of physical possibilities merely. There is no possibility of our holding an experience that directly contradicts the constitutive principles.

That is why the constitutive approach alone is unable to explain how it can be rational to abandon an established conceptual framework (a 'paradigm,' a 'research program,' a 'research tradition') in favor of a new one (Everett 2015).

II. So, Kant maintained that sublime Ideas of Reason can only play a *regulative* rather than a constitutive role. That is why they can be applied only *heuristically* as a guide for our investigations only, but not *substantially* as the actual inner principle of what we really find out. Reason governs the systematization of our knowledge: it *strives for unity*. As science advances, it subsequently replaces a 'narrower aspect of experience by a broader one.'

However, in experimental physics, – asserts Kant, - even the

> "principles according to which we perform experiments must themselves always be derived from the knowledge of nature, and hence from the theory" (quoted from Buchdahl 1969, p.510, n.1).

Thereafter, due to theory-laidenness of observations, science advances first and foremost by theory *reconciliation* and *unification*.

As a sheer representative of neo-Kantian Marburg school, Ernst Cassirer regarded that "true unity is never thought in things as such, but in *intellectual* constructions" (Cassirer [1921], 1923, p.36). In Kantian wake he stressed that unification was a purely *regulative* demand.

Along these lines, Fölsing (1997) shrewdly noted that Einstein probably first learned to think in terms of his 1905a 'heuristic viewpoint' from his early reading of Kant. Einstein's 'heuristic method' was initially to make up an assertion from which familiar facts could then be deduced. It should be pointed out that Einstein's path-breaking 1905a paper was eloquently entitled "*Über einen die Erzeugung und Verwandlung des Lichtes betreffenden* ***heuristischen*** *Lesichtspunkt*" ("On a *Heuristic Point of View* Concerning the Production and Transformation of Light").

Though throughout all his life, while carefully dissociating himself from Kantian 'synthetic a priori,' Einstein strongly supported the general neo- Kantian epistemological standpoint:

> "The following, however, appears to me to be correct in Kant's statement of the problem: in thinking we use, with a certain 'right,' concepts to which there is no access from the materials of sensory experiences" (Einstein 1944, p.22).

Thus, an important remark here is that the felicitous notion for comprehending substantial facets of Einstein's 1905 research activity is *Kant's concept of systematic Unity of Nature* as a regulative idea. This unity, both for Kant and Einstein, is not an ontological proposition at all. It is meaningless to question whether mother Nature in fact possesses such a unity or not. On the contrary, the idea of unity has *epistemological* importance first of all, since systematic unity of nature constitutes a *benchmark of validity for scientific hypothesis* that complements the empirical idea of confirmation.

Since a given scientific hypothesis cannot obtain the ultimate proof of its truth from 'below,' from repeated experiments or observations, something else is needed. One has to find out the *criterion* that can distinguish contingent and unimportant empirical generalizations from true profound Laws of Nature, which are endowed with Kantian Universality and Necessity. Kantian fruitful idea of the 'truth' of a proposition is identical to its being a law-like statement (Kant [1787], 1998, p. 592). Hence from the pile of different uniformities only those can be taken as having law-like necessity that can be fitted into a unified, systematized general system. That is why in the "*Autobiographical Notes*," ruminating on methodological principles, Einstein pointed out that

> "A system has truth-content according to the certainty and completeness of its coordination-possibility to the totality of experience. A correct proposition borrows its 'truth' from the truth-content of a system to which it belongs" (Einstein 1949a, p. 13).

It was namely this 'holistic' standpoint that allowed Einstein as early as in 1906 to call into question the widely accepted results of Kaufmann's notorious 'crucial experiments.' These data seemed to support the Abraham-Bucherer theory and to refute the "Lorentz-Einstein" one (Holton 1968, p.253; Miller 1981, p.124).

As Einstein had sagaciously put it, the rival theories (e.g. Abraham's electron theory)

> "have rather small probabilities, because their fundamental assumptions (concerning the mass of moving electrons) are not explainable in terms of theoretical systems which embrace a greater complex of phenomena" (Einstein as quoted and translated in Holton 1968, p. 253).

Einstein consented that empirically equivalent alternative theories can be easily concocted for any domain of natural phenomena. Yet if for obdurate conventionalist a la Pierre Duhem such 'an underdetermination of theory by facts' meant that there can be *no* ultimately true theory, the Einsteinian train of thought in Kantian wake not only provided a meaning to the regulative ideal of a final theory. It also highlighted Einstein's remarks that despite this underdetermination at any given time there is *only one* true theory: the theory with the *greatest power of unification* (Einstein 1918, p. 226).

Thus Einstein's enthusiastic involvement in the 1905a paper to rather dubious subject of theory of quanta was motivated by its startling *unifying possibilities*, by its fascinating capacities to arrive at an effective *fusion* of Maxwellian electrodynamics and Boltzmann's statistical thermodynamics.

Hence he commences the paper with the heart of what troubled him most – the *bottomless Abyss* in the foundations of physics felt most sharply in Lorentz's Electron Theory. *But how did Einstein strived to eliminate the basic contradiction of his 1905a paper?*

While brooding on Einstein's subtle way out of the predicament, one could not defy the fact that *all* Einstein's papers from 1901 to 1905 possess one hallmark in common: statistical-thermodynamics approach. Thomas S. Kuhn had stressed that *what prompted Einstein to idea of photon was a perseverant advancement of a research program started in 1902*, a program

> "so nearly independent of Planck that it would almost certainly have led to the black-body law even if Planck had never lived" (Kuhn 1978, p. 171).

Forsooth, just as Planck, from the outset of his startling scientific career, Einstein was fascinated by the simplicity and scope of classical thermodynamics. Though for him thermodynamics inextricably embraced the statistical approach he had imbibed from Boltzmann's aspiring writings, and so he strived to unfold *statistical* thermodynamics.

The promising result was a series of three guileless papers published in 1902, 1903, and 1904. Yet expressly *they* (see especially Einstein 1902, 1903) constitute *a steady clue* for comprehending his 1905a paper on quanta, 1905b dissertation, 1905c work on Brownian motion, and 1905d paper on STR. Just to quote the first of Einstein's three papers. This article is devoted to the foundations of statistical mechanics and was submitted to '*Annalen der Physik*' in June 1901 where his ambitious aim is boldly enunciated.

> "Great as the achievements of the kinetic theory of heat have been in the domain of gas theory, the science of mechanics has not yet been able to produce *an adequate foundation for the general theory of heat*, for one has not yet succeeded in deriving the laws of thermal equilibrium and the second law of thermodynamics using only the equations of mechanics and the probability calculus, though *Maxwell's and Boltzmann's theories* came close

to this goal. The purpose of the following considerations is *to close this gap*" (Einstein, CPAE, vol.2, Doc 3, 57; my italics).

Thus the fact that Einstein took up the blackbody problem was not accidental. He would still come to the problem because of the implementation of his own program 'even if Planck had never lived.' Besides, the first substantial result consisted in that for physical systems of jolly general kind Einstein had rendered, by the summer of 1903, both a generalized measure for temperature T and entropy S, embracing some universal constant χ.

By the time he accomplished his 1903 paper, Einstein had fathomed that χ could be properly evaluated in terms of the values of the gas constant and Avogadro's number. Yet the theory that had incited him to invoke the constant was, however, applicable to systems far more general than gases. It should, therefore, possess a correspondingly general physical foundation, engendered by a statistical-mechanical approach that led him to the constant. It soberly explained not only its lead as a scale factor for temperature but also its peculiar position as a multiplier in the probabilistic definition of entropy.

The physical significance of χ turned out to be the pivotal problem scrutinized in Einstein's third statistical paper "*On the General Molecular Theory of Heat*," submitted to the "*Annalen der Physik*" in the spring of 1904. The solution to the problem was inextricably connected with the queer phenomena of energy fluctuations. Einstein expounded that

$$\bar{\varepsilon} = 2\chi \text{ T dE/dT},$$

where $\bar{\varepsilon}$ is a measure of thermal stability of the system, T the temperature of the system, and E its energy. Moreover, it was comprehension of the constant physical sense that directed his attention to the blackbody problem.

"The equation just found would permit an exact determination of the universal constant χ if it were possible to determine the energy fluctuation of the system. In the present state of our knowledge, however, that is not the case. Indeed, *for only one sort* of physical system can we presume from experience that an energy fluctuation occurs. That system is empty space filled with thermal radiation" (Einstein 1904, p.360, my italics; translated in Kuhn 1978).

In 1904 at least one more step in the advancement of the program of statistical thermodynamics had to be taken, and Einstein resolutely took it in the subsequent 1905a paper. Its content reveals that Einstein had commenced to look for a black-body law of his own, and that he had quickly excavated the paradox, evinced in the contradiction between statistical mechanics and Maxwellian electrodynamics. Ipso facto, he had dropped the search for the law in favour of a survey of the paradox itself. It is clear from the beginning of his already quoted article (translated in Ter Haar 1967). The first part of the 1905a 'photon' paper came to an end by eliciting the notorious 'ultraviolet catastrophe.' *But how did Einstein intended to resolve the stodgy paradox?*

In the second part of his 1905a paper Einstein brings together thermodynamics, statistical mechanics and Maxwellian electrodynamics to survey the domain of empirical reality covered by Wien's radiation law. The joint application of the three mature theories, belonging to diverse discrete research traditions of classical physics enables Einstein to draw the following conclusion: if monochromatic radiation of frequency ν and energy E is enclosed in the volume V_0, then the probability W that at any moment all the radiation energy will be discovered in the partial volume V of the volume V_0 is given by

$$W = (V/V_0)^{E/h\nu} \qquad \text{(i)}$$

Yet in the same paper Einstein had previously ascertained that in the case of n independently moving particles enclosed in a volume V_0 the probability of finding them all momentarily in the subvolume V is

$$W = (V/V_0)^n \qquad \text{(ii)}$$

Comparing equations (i) and (ii), Einstein draws a startling conclusion that

> "monochromatic radiation of small density behaves in thermodynamic respects *as though it consists of distinct independent energy quanta of magnitude hν.*"

Note that the odd conclusion that radiation in the cavity 'consists' of independent energy quanta follows *directly* from application of general principles of thermodynamics and statistical mechanics to radiation phenomena.

Nevertheless, at the end of the XIX-th century all the available experimental data, relevant to fluorescence, photoelectricity and photoionization data, could warrant only *indirect* verity of quantum hypothesis. Thus, to rigorously test the 'ultra-revolutionary,' bizarre hypothesis of quanta, Einstein had to appeal to a '*crucial experiment*' of a very peculiar, freaky kind. He had to match the quantum results with the results of another entrenched, 'old' theory contrived *independently* of the 1905a thrilling 'photon' hypothesis.

This theory had to be sufficiently 'old' to accumulate the results of various reliable experiments. And if the 1905a paper consequences had matched those of fairly different theory, that arose out of substantially different problem situation, they would have provided an especially reliable verification of dizzying 'photon hypothesis.'

In the opposite case the 1905a theory could have been 'falsified' not by a single 'critical experiment' but by a whole multitude of the well-established experimental data. Surely it was this 'holistic' stand that allowed Einstein as early as in 1906 to disregard the results of Kaufmann's 'crucial' experiments, which seemed to buttress the Abraham-Bucherer theory and to refute the "Lorentz-Einstein" theory (Holton 1968, p.253; Miller 1981, p.124).

Indeed, as Einstein had recalled later, the rival theories (e.g. Abraham's electron theory)

> "Have rather small probabilities, because their fundamental assumptions (concerning the mass of moving electrons) are not explainable in terms of [mature] theoretical systems which embrace a greater complex of phenomena" (Einstein as quoted in Holton 1968, p.253).

Thus the next - 1905b – 'Brown motion' result turned out crucial for the 1905a verification. In the subsequent 1905b paper Einstein sedulously perpetrated the principles of Brownian motion that were directly verified by Perrin's famous experiments.

> "My principal aim in this [1905b work on Brownian motion] was to find facts that would guarantee as much as possible the existence of atoms of definite size... The agreement of these considerations with experience together with Planck's determination of the true molecular size from the law of radiation (for high temperatures) convinced the sceptics, who were quite numerous at that time (Ostwald, *Mach*), of the *reality of atoms*" (Einstein 1949a, pp.45-47; my italics).

Regrettably, the importance of 1905b paper's for the 1905a one was disclosed by Einstein much later. He admitted to Max von Laue on 17 January 1952:

> "When one goes through your collection of verifications of the special relativity theory, one believes that Maxwell's theory is firmly established. But in 1905 I knew *already with certainty* that it leads to the wrong fluctuations in radiation pressure, and consequently to an incorrect Brownian motion of a mirror in a Planckian radiation cavity" (quoted from Rynasiewicz 2000, p.177; my italics).

This conspicuous for 1905 Einstein result was presented to the scientific community only in 1909 when Einstein masterly applied his theory of Brownian motion to a two-sided mirror immersed in thermal radiation. He disclosed that the mirror would be unable to carry out a Brownian motion indefinitely, if the fluctuations in the radiation pressure on its surfaces were solely due to the effects of random waves, as asserted by Maxwell's theory. Yet only the presence of an *additional* term, corresponding to pressure fluctuations due to the impact of random particles, insures the continued Brownian motion of the mirror.

Einstein revealed that similar fluctuation terms in the energy were necessary consequences of Planck's law. He considered such fluctuation phenomena as the *strongest* argument for ascribing physical significance to the hypothetical light quanta (Stachel 2000; Bordoni 2013). Only after this queer 'crucial thought experiment,' that is only *after* the 1905b paper could Einstein look forward for investigating the startling consequences of his light quantum hypothesis. And so he returned to his half-forgotten 'unsere Arbeit uber die Relativbewegung,' his former 'kapitale Abhandlung.' Namely this protracted humble manuscript was destined to become the celebrated 'special theory of relativity.'

So far, so good.

> "If the monochromatic radiation (of sufficiently small density) in the sense of entropy dependence upon volume behaves itself as a discontinuous medium, consisting of energy quanta Rβν/N, a question occurs: if they are not the laws of creation and conversion of light such as if it consists of similar energy quanta?" (Einstein 1905a, p. 236).

That is the question cautiously put up by Einstein at the end of § 6 of his 1905a. But it was *the ether conception that turned out to be a substantial stumbling block.* It strongly prevented positive answer and put insurmountable obstacles in exerting Einstein's statistical-thermodynamics program. Indeed,

> "mechanical and purely electromagnetic interpretations of optical and electromagnetic phenomena have in common that in both cases electromagnetic field is considered as a special state of the hypothetical medium filling all the space. Namely in that point two interpretations mentioned *differ radically* from Newton's emission theory, in which light consists of moving particles. According to Newton, space should be considered as possessing neither ponderable matter, nor light rays, i.e. absolutely empty" (Einstein 1905a, p.236, my italics).

To work out a quantum theory of radiation, one needs electromagnetic fields as substantially *independent* entities that can be emitted by the source 'just as in Newton's emitting theory.' (Viz., the energy transmitted in a process of emission should not be dissipated in space, but should be *completely* preserved until an act of absorption).

Yet within the Lorentz programme an electromagnetic field is taken as a specific state of ether - a state of medium that is *continuously* distributed in space. In such a medium an elementary process of radiation is connected only with a *spherical* wave.

Rather, Einstein (1905a, p.237) never thought of disposing Maxwell's theory, just as Newton, the inventor of the emission theory, did not reject the wave theory 300 years earlier. In Lorentz's splendid theory this stumbling block was absent. Certainly, in the reference frame that is at rest relative to the ether light propagates with constant velocity *independent of the velocity of the source.* Thus, if one wants to give the idea of ether up, but to come to terms with Maxwell's theory at the same time, s/he should disown ballistic hypothesis and

> "raise to the rank of a principle the validity of the law of constancy of light velocity for all inertial frames" (Einstein 1936 /1954, p.307),

i.e. *postulate a special "principle of constancy of velocity of light" (I).*

The second basic STR postulate - *"the principle of relativity" (II)* - follows directly from the abovementioned premise that there is no luminiferous ether and, consequently, no absolute system of reference. Just to recall the "*Autobiographical Notes*,"

> "Reflections of *this type* [i.e. on molecular structure of radiation] made it clear to me as long as shortly after 1900, i.e. shortly after Planck's trailblazing work, that neither mechanics nor electrodynamics could (except in limiting cases) claim exact validity. By and by I despaired of the possibility of discovering the true laws by means of constructive efforts based on known facts. The longer and the more despairingly I tried, the more I came to the conviction that only the discovery of a *universal formal principle* could lead us to assured results. *The example I saw before me was thermodynamics*" (Einstein 1949a, p.51; my italics).

The latter point needs further elucidation by delving into Mach's pre-eminent "*Mechanics*" as the main source of 1905 Einstein's knowledge on the history and methodology of physics. The most thorough case study of the interconnection between the principle of economy of thought and second law of thermodynamics in "*Mechanics*" is Stevinus's (1548-1620) felicitous theoretical scheme of statics. In his immaculate "*Hypomnemata Mathematica*" (Leyden 1605) Stevinus pioneered in eliciting the mechanical properties of the inclined plane. He strived to ascertain a general theoretical principle and then to investigate the partial cases that can be easily treated by quantitative means. To proffer the pivotal gedankenexperiment, necessary to set up his general principle, Stevin contrives a triangular prism with no horizontally placed edges. Over the prism an endless string is laid on which 14 balls of equal weight are strung and tied at equal distances apart. (The string can be easily replaced by an endless uniform chain).

This chain will either be in equilibrium or not. If we presume the latter to be the case, the chain, since the conditions of the event are not altered by its motion, must, when once actually in motion, continue to move forever. So, it must set up perpetual motion, which Stevin takes blatantly absurd. Thus only the first case is conceivable and *the chain always remains in equilibrium.*

According to Ernst Mach, there is contained only a *purely instinctive* cognition in the basic premise from which Stevin starts (that the endless chain does not move). He feels at once, and we with him, that we have never observed anything like a motion of the bizarre kind referred to. This conviction has much logical cogency.

Hence one eagerly accepts the conclusion drawn from it respecting the law of equilibrium on the inclined plane without the thought of an objection. Although the law is shrewdly presented *as the simple result of the experiment.* We cannot be surprised at this when we take into account that all results of experiment are obscured by adventitious circumstances (as friction, etc.). And that every conjecture as to the conditions which are determinative in a given case are liable to error. Thereby Stevinus ascribes to instinctive knowledge of this sort a *higher authority* than to simple, apparent, and direct observations!

So, we are faced with the following substantial question: w*hence does this higher authority come?* If one recalls that scientific demonstration, and scientific criticism generally can only have sprung from the consciousness of the individual fallibility of investigators, the explanation is apparent. We feel clearly, that we ourselves have contributed *nothing* to the creation of this miraculous '*Instinctive Knowledge*.' And that we have added to it nothing arbitrarily, but that it exists in strict independence of our participation.

According to Mach, Stevinus's deduction is one of the most unique 'fossile indicators' that we possess in the history of mechanics. It throws a fascinating light on the process of the genesis of scientific concepts generally, on their bland genesis from instinctive knowledge.

Yet, every experimenter can daily underscore in his own person the guidance that Instinctive Knowledge furnishes him. If he succeeds in *abstractly formulating* what is contained in it, he will as a rule have made a significant advance in science. And it is perfectly certain for the author of "*Mechanics*": it is the amalgamation of the strongest instinct with the greatest power of abstract formulation that constitutes the great natural inquirer (Mach [1893], 1999: p.27).

Though how does this 'instinctive knowledge' engender and what is its subtle structure? Everything which one contemplates in nature imprints itself *uncomprehended* and *unanalysed* in his percepts and ideas. In these accumulated experiences he possesses a 'treasure store' which is ever close at hand and of which only the small droplet is embodied in fine articulate thought. The circumstance that it is far easier to appeal to these experiences than it is to Nature herself, and they are, notwithstanding this, free from all subjectivity, *invests them with such a high value*.

> "It is a peculiar property of instinctive knowledge that it is predominantly *of a negative nature*" (Mach [1893], 1999, p.28].

We cannot so well state what must happen as we can what *cannot* happen, since the latter alone stands in devastating contrast to the obscure mass of experience in us in which single characters are not distinguished. Moreover, contends Mach, the other hallmark is no less substantial for the epistemology.

The sober reasoning of Stevinus has such a strong influence upon us because the result at which he arrives obviously contains *more* than the assumption from which he starts.

Nevertheless, it often happens in the course of the advancement of science that a new principle perceived by some researcher in connection with a fact, is not immediately recognized and set up in all its generosity. If, throughout all available facts, we clearly indicate and *discern* a principle which, though not admitting of proof, can yet be known to dominate, we have advanced much farther in the consistent view of nature than if we suffered ourselves to be overawed by a specious demonstration (Mach [1893], 1999, p. 82).

All in all, in my view, *all the abovementioned Stevin-Mach fascinating recipes where ingeniously implemented by Einstein in formulating the fundamental STR principle – the principle of relativity.* Though due to ultra-revolutionary and pronouncedly speculative nature of light-quanta hypothesis he could not dare to disclose the link with the 1905a paper straightforwardly. Hence he applied all the Stevin-Mach handy technique of conviction to present his 'electrodynamics of moving bodies' in purely *phenomenological wake.* One should especially pay respect to the *negative character of the relativity principle* and the peculiar manner of its connections with experiments and observations. It is much closer to subtle conviction technique of instinctive knowledge than to blunt inductive way of inference. Behold the beginning of the STR paper:

> "Examples of a similar kind, and the failure of attempts to detect a motion of the earth relative to the 'light medium,' lead to the *conjecture* that not only in mechanics, but in electrodynamics as well, the phenomena *do not* have any properties corresponding to the concept of absolute rest, but that in all coordinate systems in which the mechanical equations are valid, also the same electrodynamic and optical laws are valid, as have already been shown for quantities of the first order. *We shall raise this conjecture* (whose content will be called 'the principle of relativity' hereafter) *to the status of a postulate* and shall introduce, in addition, the postulate, only seemingly incompatible with the former one, that in empty space light is always propagated with a definite velocity V which is independent of the

state of motion of emitting body" (Einstein 1905d, p.140; my italics).

An eloquent appeal to *instinctive knowledge* easily explains the 'hard fact' that the STR paper by rights stands out in all the world scientific literature for the *complete lack of quotations*. Just to recall Hendrik Lorentz's 1906 lecture at Columbia university:

> "His [Einstein's] results about electro-magnetic and optical phenomena ... are in general in agreement with those we have obtained in the previous pages, the principal difference being that Einstein simply *assumes* what we have deduced, with some difficulties and not always in satisfactory manner, from the fundamental equations of the electro-magnetic field. In doing this, he can surely be credited with the merit of showing in the negative result of experiments like those of Michelson, Rayleigh and Brace, not a fortuitous compensation of contrasting effects, but the manifestation of a general and fundamental principle" (quoted from Baracca, 2005, p.22).

Furthermore, since, according to the "*Autobiographical Notes*," Einstein's new hybrid theory was contrived as a result of inevitable encounter of Newtonian mechanics and Maxwellian electrodynamics, its basis should consist of a *minimum* of two postulates. (I) The first one drawn from classical mechanics (the principle of relativity) and (II) the second one adroitly transferred from the Maxwell-Lorentz electrodynamics (the principle of the constancy of light). Namely,

> (I) "*Classical mechanics*, of which it could not be doubted that it holds with a close degree of approximation, teaches the equivalence of all inertial systems or inertial 'spaces' for the formulation of natural laws, i.e., the invariance of natural laws with respect to the transition from one inertial system to another" (Einstein 1954, p.369).
>
> (II) "This [the special theory of relativity] takes over from the *theory of Maxwell-Lorentz* the assumption of the constancy of the velocity of light" (Einstein 1940 /1954, p. 370).

The two postulates, (I) + (II), the relativity principle in conjunction with the principle of constancy of velocity of light, are quite sufficient, according to Einstein, to deploy the electrodynamics of moving bodies. Yet, since

> "the theory based on these two principles should not to lead to contradictory results, one must renounce the customary rule of addition of velocities" (Einstein 1910, p.125).

And namely that was exhibited in the pre-eminent 1905d paper "*On the Electrodynamics of Moving Bodies*," published only several months *after* the 1905a one. Einstein had revealed the hidden assumption - the basis of the Galileo addition law - that the statements of time, as well as of the shapes of moving bodies have the sense independent of the state of motion of the reference frame. He elucidated (together with Michele Besso) that the acceptance of the 'principle of relativity' in conjunction with the 'principle of constancy of light' is equivalent to significant modification of the simultaneity concept and to clock delay in moving reference frame.

In a sense, one can conclude, with S. Abiko (1991, 2005) that the STR 'derives from chemico-thermal studies.' However, Einstein sagaciously did not promulgate the connections between 1905a and 1905d until 1909, but had to bare the subtle link between the 1905a and 1905d startling innovations four years later. In 1909, in Salzburg, he made a historical overview report at the 81-st meeting of German Natural Scientists and Physicians promisingly self-titled "*On the Development of our Views on the Nature and Structure of Radiation*."

It was practically the first effort and opportunity to exhibit almost all his various papers as a whole. And it represented one of the first Einstein's public reports dedicated to expounding of the STR discrete foundations. The report commences with a brief recapitulation of luminiferous ether theory that ends by an intriguing question: "However, today we must take the ether hypothesis as an obsolete standpoint." Why? What I want to punctuate is that for the answer Einstein appeals *not* to the Michelson-Morley or Fizeau experiments, but stresses that

> "It is even undeniable that there is an extensive group of facts concerning radiation that shows that light possesses certain fundamental properties that can be understood far more readily from the standpoint of Newton's *emission theory of light* than from the standpoint of the wave theory. It is therefore my opinion that the next stage in the development of theoretical physics will bring us a theory of light that can be understood as a kind of *fusion* of the wave and emission theories of light" (Einstein 1909, p. 379; my italics).

To sum up, the Planck-Einstein breakthrough, or the quantum-relativistic revolution can be described as a 'Hacking-type' (Schweber 2017) revolution, named after Ian Hacking. The probabilistic revolution of the nineteenth century was characterized (Hacking 1987) in terms of the change of 'style of reasoning.' A style of reasoning embraces types of objects, evidence, sentences, ways of qualifying truth or falsehood, laws, modalities and most importantly, new possibilities. The hallmark of Hacking-type revolutions consists in that they are 'emplacement revolutions,' rather than replacement ones. Viz., they change the way science is *practiced* without decisive abandoning all the previous concepts by transforming it from within.

CHAPTER V.

Paul Dirac's Distinctive Synthesis of Quantum Mechanics and Special Relativity: A Trace of Transcendental Idealism.

A theory with mathematical beauty is more likely to be correct than an ugly one that fits some experimental data
— **Paul Dirac**

My work has always tried to unite the truth with the beautiful, but when I had to choose one or the other, I usually chose the beautiful
— **Hermann Weyl**

Abstract. One of the odd points of intricate history of XXth century physics – Paul Dirac's stupendous contrivance of the relativistic theory of the electron – is elucidated in the context of apt mature theory- unification model. The rational reconstruction of the Dirac synthesis of special relativity and quantum mechanics is proffered that makes it possible to compare it with the classical synthetic samples (Newton, Maxwell, Einstein) and to disclose its specificity. The fine structure of Dirac's intricate research program and first and foremost the odd principles that put up its powerful heuristics is scrutinized with special emphasis on the highly controversial tenets of 'mathematical beauty' and 'unity of science.' Digging still deeper into the background, it is contended that despite the relentless Dirac's remarks denigrating the role of philosophy one can trace a tinge of its indirect influence epitomized in Husserlian doctrine through Arthur Eddington's and Hermann Weyl's whimsical mathematical writings embracing the pioneering efforts to merge gravitation with electromagnetism. Accordingly, the milestones of Dirac's research program realization in the distinctive context of the applied model are elicited.

Keywords: quantum mechanics, special relativity, synthesis, Dirac, Eddington, Einstein, Weyl, Husserl.

V.1. Blunt introduction.

Paul Dirac's startling amalgamation of the special theory of relativity (STR) and nonrelativistic quantum mechanics (NQM) occupies a peculiar place in the history & philosophy of science due to the following apparent circumstances.

First, Dirac's thought-provoking relativistic theory of the electron played a particularly significant asset in the second scientific revolution in general and in the contrivance of both QED (quantum electrodynamics) and QFT (quantum field theory) in particular.

Second, Dirac's skeptical attitude to epistemology is too well-known to dwell on. It therefore comes as no surprise that the author of this chapter cannot recall neither the name of the renowned philosopher, nor the name of the philosophical school, to which Dirac definitely referred at least once, not to mention a single humble quotation from *any* philosophical writing. Moreover, Dirac himself adamantly avoided philosophical discussions. For instance, even despite his unorthodox views, he never evinced any public interest in the opposition waged against Bohr's 'Copenhagen weltanschauung' by Einstein, Schrödinger, de Broglie, and other pundits.

A certified electrical engineer, a guileless graduate of the University of Bristol, was characterized by a markedly caustic attitude towards philosophy, which tried hard yet was not able to ascertain the definite and stout guidelines necessary for successful research activity.

> "I feel that philosophy will *never* lead to important discoveries. It's just a way of talking about discoveries which have *already* been made" (Dirac interview, 1963, quoted from Kragh, 1990).

In this subtle respect, Dirac's views were blatantly at odds with the Weltanschauung of another smart youngster and one of his strongest competitors - Werner Heisenberg, the future author of thought-provoking '*Physik und Philosophie*' (2000), who often and willingly quoted Immanuel Kant and participated in Martin Heidegger's memorial volumes. Or, say, Niels Bohr himself who reflected even on Sören Kierkegaard's sad fate. (Not to slight thoughtful Erwin Schrödinger, who craved for teaching philosophy professionally).

To this one should add the almost complete absence of notebooks, letters, diary entries, eyewitness accounts, etc., which would make it possible to elicit with confidence Dirac's alluring '*logic of discovery*.' Forsooth, fortunately we have a soothing Dirac's indication that the 'true history of his research does not deviate too much' from that which can be reconstructed based on an analysis of his articles (Kragh 1981).

Nevertheless, all of the above not only does not eliminate but on the contrary - conditions - the need for a rational reconstruction of the Dirac unification of SRT and nonrelativistic QM, especially for comparing it with classical unification samples (Newton, Maxwell, Einstein) and unfolding its originality. The first important difference that immediately catches the eye is the presence of apparent *philosophical* premises in the abovementioned 'classical' group. And it's not surprising since the unification of quite distinct fields of inquiry necessitates an abstract philosophical outlook and reflection on methodological principles

One cannot flinch from asserting a blunt theological influence in Newton's renowned case. Before the austere author of '*Principia*,' Aristotelian – Ptolemaic pagan cosmology blatantly confronted the stiff principles of monotheism not admitting the blunt and impervious demarcation line between the celestial and mundane antagonistic realms. All the adverse worlds should have the same Creator (Nugayev 2023) 'an impressive Pantocrator in the dome.' Whereupon, "God said 'Let the Newton be' and all was light" (Alexander Pope).

In Maxwell's distinctive case the pervasive influence of Kantian epistemological doctrine providing the inexorable unification of diverse electrical and magnetic mathematical models turned out to be indispensable.

> "in the meantime, I have my usual superfluity of plans…4. Metaphysics – Kant's Kritik of Pure reason in German, read with a determination to make it agree with Sir W. Hamilton…" (Campbell &Garnett 1882).

Forsooth, Maxwellian electrodynamics was contrived as a result of the old pre-Maxwellian programs' reconciliation: the stringent electrodynamics of Ampère-Weber, the profound wave theory of Young-Fresnel and Faraday's queer program. Eventually, Maxwell's victorious program pressed out the Ampère-Weber one because it did assimilate the instant-action ideas, as well as the robust presuppositions of the programs of Young-Fresnel and Faraday. Maxwell's landslide victory became possible because the core of his unification strategy was permeated by Kantian epistemology. Viz., Maxwell did put forward as a pivotal synthetic principle the idea that radically differed from that of rival standpoints by its open, flexible, and contra-ontological character. 'Action at a distance,' 'incompressible fluid,' 'molecular vortices' were merely 'contrived analogies' for Maxwell, capable only to direct the researcher at the 'right' mathematical relations (Nugayev 2015).

In Einstein's inextricable case the latent influence of Kantian epistemology was mediated and amended by new-fangled writings of Mach and Duhem. Forsooth, to comprehend in what subtle ways Einstein's medley 1905 breakthrough papers hang together one has to pay special tribute to his celebrated *strive for unity.* Einstein's quirk light quanta and terse special relativity turn out to be only pencil sketches over a more profound oil painting, mere stages of unwinding of Maxwellian electrodynamics and statistical thermodynamics reconcilement research program. The stale conception of luminiferous ether was a significant snag for Einstein's statistical thermodynamics in which the predominant role was played by the whimsical light quanta paper. Whereupon Einstein was fully aware that his light quanta hypothesis was too audacious to be taken literally. Hence his 'electrodynamics of moving bodies' was posited in a markedly Machian, phenomenological way, and the basic relativity postulate was proposed as organically originating from the wayward 'instinctive knowledge.' In his relentless criticism of the mold research traditions of classical physics Einstein was famously influenced by David Hume and Ernst Mach. Nevertheless, when related to creative momenta, his 1905 reconcilement *modus operandi* was initiated by Mach's pre-eminent principle of economy of thought though taken in the context of

Stevin-Mach 'instinctive knowledge' doctrine altogether with inclinations of the regulative spirit of Kantian epistemology (Nugayev 2019).

So, what was Dirac's own philosophy of science that allowed him to unify such diverse realms as 'relativity' and 'quanta'? Regrettably, Dirac's implicit philosophy of science has to be piecemeally reconstructed from his stout papers and scant writings on general questions. It is one of the aims of this chapter to contend that, despite the relentless Dirac's remarks on the notorious role of philosophy, one can trace a latent influence of Husserlian epistemology through Arthur Eddington's and Hermann Weyl's whimsical mathematical writings embracing their paradigmatic efforts to merge gravitation with electromagnetism. It is not an accident that Dirac overtly characterized the defunct Weyl theory as "a ***very beautiful synthesis*** of the electromagnetic field and the gravitational field," a theory that "remains as the outstanding one, **unrivaled** by its *simplicity and beauty*" (Dirac 1973a, p.53). It is not an accident that Dirac's renowned "*The Principles of the Quantum Mechanics*" was praised by an anonymous reviewer in "*Nature*" (most probably by Eddington) as "the application of *pure thought* to physics." And" it is this that makes Dirac's method more *profound* than that of other writers" (Kragh 1990). To furnish the process, the chain of inextricable interactions, comprising such important links as "Weyl-Eddington," "Eddington-Dirac," "Weyl-Dirac," and finally "Husserl-Weyl" will be envisaged.

Hence, the overall *aim* of this chapter is to try, despite the obvious absence of *direct* instructions from Dirac, to highlight the basic heuristic principles of the synthesis of SRT and nonrelativistic NQM in their relationship with transcendental idealism (first section) and layout how they were distinctively implemented in actual scientific practice (second section).

V.2. The Pivotal Heuristic Principles of Dirac's Distinctive Scientific Research Program and Their Epistemological Grounding.

In my view, even though the existence and functioning of a conscious metaphysical 'hard core' is extremely doubtful, one can still admit the existence of the 'positive heuristic' of Dirac's SRP (scientific research program), which, without any significant changes, permeated *all* his scientific creativity.

(I) As the *first principle*, one can single out Dirac's blunt and solid beliefs about the smooth, evolutionary development of science and the need for prudent and gradual (piecemeal) changes in the content of mature scientific theories (Bokulich 2004). In this respect, Paul Dirac's unobtrusive views were definitely at odds with those of his celebrated friend, colleague, and rival, Werner Heisenberg. The latter advocated a *holistic* interpretation of the physical theory with enthusiasm. The standpoint consisted, first of all, in a subtle Duhemian thesis on the integrity of the content of a scientific theory, when the slightest changes in its content in a discreet place instantly led to corresponding changes in all the nooks. Accordingly, a scientific revolution incarnates a simultaneous (and abrupt) change of the entire 'paradigm' at once. Incidentally, one of the important consequences of this circumstance is the renowned 'Kuhn-Feyerabend thesis' on the incommensurability of the contents of the 'old' and 'new' mellow paradigms.

Dirac, starting from his early 'postgraduate' articles, constantly - and until the end of his life - stressed the *continuity* of QM and classical mechanics (Dirac 1927; Dirac 1938-39) in accordance with domestic research traditions.

(II) *The Principle of Unity of Science* asserted that "all of Nature is amenable to the same kind of theoretical treatment" (Kragh 1990, p.267) Forsooth, in the long run, it is counter-productive to operate with multilateral and frequently incompatible theoretical schemes in different realms of Nature.

For instance, commenting acrimoniously on the bootstrap controversy in 1966, Dirac asserted:

> "In physics, one should aim at a *comprehensive* scheme for the description of the *whole* of Nature. A vast domain in physics can be successfully described in terms of equations of motion. Quantum field theory must be based on concepts and methods that

can be *unified* with those used in the rest of physics" (Dirac 1969, p.4).

Three years later Dirac related his reasons against the S-matrix or bootstrap approach more concisely:

> "The reason for this disbelief comes essentially from my belief in the unity of the whole physics. High energy physics forms only a very small part of the whole of physics- solid-state physics, spectroscopy, the theory of bound states, atoms and molecules interacting with each other, and chemistry. All these subjects form a domain vastly greater than high-energy physics, and all of them are based on the equations of motion. We have reason to think of physics as a whole and we need to have the same underlying basis for the whole of physical theory" (Dirac1969, p.4).

(III) The *third* in a row (but not in importance!) - put forward by Dirac as a *pivot* - the so-called '*principle of mathematical beauty.*' When he once again visited Moscow in 1956, he famously inscribed on the ceremonial Lomonosov State University blackboard: "A physical law must possess mathematical beauty." It is a platitude that no other outstanding physicist had been preoccupied with the enticing concept of beauty as was Paul Dirac. This principle was so dear to the QED father that

> "A theory with mathematical beauty is more likely to be correct than an ugly one that fits some experimental data" (Dirac 1970; quoted from Bokulich 2004).

Incidentally, 24-years-old Dirac had begun his Easter term of 1926 Cambridge lecture course, which turned out to be the first course ever on QM at a British university (and attended by, among others, Nevil Mott and Julius Robert Oppenheimer), indicating the stale empirical inadequacy of classical theories of physics, but quickly turned on defiantly aesthetical ground:

> "The chief thing that the quantum theory had to fight against is the fact that classical electrodynamics is a very *elegant* complete self-

> consistent theory based on a few simple assumptions, and it appears that any modification of such a theory must be *ugly,* and not what we would expect nature to be like" (quoted from Ryckman 2005, p.250; my italics).

(Note that the content of Dirac's early lectures formed the ground of his unexceptionable "*The Principles of Quantum Mechanics*" published in summer 1930).

Tireless and inquisitive researcher, who wrote his last research paper at the age of 77, till his last days was faithful to the aesthetically founded conviction. He complained to Dyson: "I might have thought that the new ideas [QED] were correct if they had not been so *ugly*."

Yet it is promising (for future layout) to compare these quotes with Weyl's significant revelation:

> "My work has always tried to unite the truth with the beautiful, but when I had to choose one or the other, I usually chose the beautiful."

Regrettably, one cannot find in Dirac's writings the definition of the criterion of 'mathematical beauty.' For instance, in a talk given at the University of Miami in 1972, entitled "*Basic Belief and Fundamental Research,*" Dirac is obliged to admit that

> "mathematical beauty is of a rather different kind [in comparison with literature and music]. I should say perhaps it is of a completely different kind and transcends these personal factors. It is the same in all countries and at all periods of time" (Ryckman 2005, p.288).

(IV) The *fourth,* all appearance, no less significant principle of the Dirac heuristic is closely related to those considered above: they are Einstein's theories of relativity (STR and especially GTR) that constitute a sublime *ideal*, and unsurpassed specimen of mathematical beauty. First,

> "The Lorentz transformations are beautiful transformations from the mathematical point of view, and Einstein introduced the idea

that something which is beautiful is very likely to be valuable in describing the fundamental physics" (Dirac 1980)

Second,

"When Einstein was working on building up his theory of gravitation he was not trying to account for some results of observation. Far from it. His entire procedure was to search for a beautiful theory, a theory of a type that nature would choose. He was guided only by the requirement that the beauty and elegance which one would expect to be provided by any fundamental description of nature."

Third, writing back to Werner Heisenberg in response to his manuscript Dirac stressed:

"My main objection to your work is that I do not think your basic (non-linear field) equation has sufficient *mathematical beauty* to be a fundamental equation of physics. The correct equation, when it is discovered, will probably involve some *new* kind of mathematics and will excite great interest among the pure mathematicians, just like Einstein's theory of the gravitational field did and still does" (Dirac to Heisenberg, March 6, 1967, quoted from Brown & Rechenberg 1987, p.148).

Whereupon, Einstein's so-called 'geometrization of gravitational interactions' in 1915 proffered the modern geometrical unification program in physics. Accordingly, Einstein's unique experience of constructing the GTR was of decisive importance for the problem of relativistic electron contrivance and hence for proper combining STR and nonrelativistic QM. Especially strongly, according to the abovementioned quotation, Dirac was staggered by the dazzling fact that the GTR was created *without any apparent connection with experiment*, and, as Arthur Eddington never tired of emphasizing, *by no means* as a direct generalization of experimental data. In this regard one should recollect Eddington's notorious claim that the GTR basic equations are 'mere definitions.' Yet even this outrageous conclusion can be defended, according to Ryckman, by pointing to the idiosyncratic way of expressing the famous Kantian dictum that “Reason has insight only into that which it produces after a plan of its own.”

It is not an accident that in his Herbert Spencer lecture of 1933 Einstein famously maintained:

> “Our experience hitherto justifies us in believing that Nature is the realization of the simplest conceivable mathematical ideas. I am convinced that we can discover using purely mathematical constructions the concepts and the laws connecting them, which furnish the key to the understanding of natural phenomena. Experience may suggest the appropriate mathematical concepts, but they most certainly cannot be deduced from it. Experience remains, of course, the sole criterion of the physical utility of mathematical construction. *But the creative principle resides in mathematics*. In a certain sense, therefore, I hold it true that pure thought can grasp reality, as the ancients dreamed” (Einstein 1933, pp. 17-18, my italics).

Accordingly, the GTR not only earnestly explained the anomalous precession of the perihelion of Mercury, known for more than half a century, but also amazingly predicted the unrehearsed deflection of light rays in the gravitational field of the Sun. The bold prediction was famously confirmed in 1919 by an expedition of British astronomers again led by Sir Arthur Eddington.

Extremely important for the professional formation of the English theoretical physicist was the fact that from the very beginning of his acquaintance with the GTR, Dirac was so fascinated by it (partly thanks to the influence of the writings, lectures, and discussions with Sir Arthur) that he dreamed of studying this very theory in Cambridge. And only when it became clear that the corresponding tutor was absent there (since Dr. Ebenezer Cunningham refused to be a supervisor), the former unemployed electrical engineer was forced to accept Dr. Ralph Fowler’s magnanimous offer to dwell into the bizarre realm of quantum theory. But even in the heavy process of preparing for his Ph.D., Dirac could not deny himself the pleasure of attending the fascinating lectures on STR, GTR, and tensor analysis, delivered by the greatest English expert in the field, prof. Arthur

Eddington. (Not by chance Einstein considered Eddington's "*The Mathematical Theory of Relativity*" as "the finest presentation of the subject in *any* language"). Moreover, Dirac had the lucky opportunity to thoroughly and thoroughly discuss its foundations in personal meetings.

(V) Hence no less important, from my point of view, is the hard fact that the 'father of relativistic electron' got acquainted with the GTR not, so to speak, 'in the original,' but only *through* the prism of the distinctive writings of Sir Arthur Eddington, who, as 'the Apostle of Relativity in Britain,' was known in the post-war England 'even more than Einstein himself' (Dirac).

> "We had no chance to understand relativity *properly* until 1923 when Eddington published his book '*The Mathematical Theory of Relativity*' which contained *all* the information needed for a *proper* understanding of the basis of the theory. This mathematical information was interspersed with *a lot of philosophy*. Eddington had his philosophical views, which, I *believe,* were somewhat *different* from Einstein's, but developed from them" (Dirac; quoted from Ryckman 2005, p.182; my italics).

An intriguing quote. What kind of queer Eddington's 'philosophy of relativity' the no less eccentric creator of the theory of the relativistic electron had in mind? – Maybe he meant the famous Eddington dictum that "the idealistic tinge in my conception of the physical world arose out of mathematical researches on the relativity theory" (Eddington [1928], 1948, p. vii)? Or, maybe, no less celebrated, another one: "Science aims a constructing a world which shall be *symbolic* of the world of commonplace experience" (Eddington [1928], 1948, p. vii)? Incidentally, to Herbert Dingle, A. Eddington and P.A.M. Dirac transpired as 'modern Aristotelians,' worshipping at the idol of a rationalistic cosmophysics, the arch-practitioners of a dangerous rite posing the "radical danger" to the foundations of the science of prostrating observation before the golden calf of mathematical invention" (Ryckman 2005, p.176).

A closer look at the preface of "*The Principles of Quantum Mechanics*," first published in 1930, elicits the common ground between the two scholars. After he had underscored the importance of the theory of transformations, Dirac, directly 'borrowing phrases from Eddington's philosophy' (Kragh), notes:

> "[Nature's] fundamental laws do not govern the world as it appears in our *mental picture* in any very direct way, but instead they control a substratum of which we cannot form a mental picture without introducing *irrelevancies.* The formulation of these laws requires the use of mathematics of *transformations.* The important things in the world appear as ***the invariants*** (or more generally the nearly invariants, or quantities with simple transformation properties) of these transformations...The growth of the use of transformation theory, as applied first to relativity and later to the quantum theory, is ***the essence*** of the new method in theoretical physics. Further progress lies in the direction of making our equations invariant under wider and still wider transformations. This state of affairs is very satisfactory from a ***philosophical point of view***, as implying an increasing recognition of the part played by the observer *in himself* introducing the regularities that appear in his observations, and a lack of arbitrariness in the ways of nature" (Dirac [1930], 1981, pp. iv-v; my italics).

Another important (and single!) quote from Dirac can be also engaged. Indeed, back in 1931, Dirac prefaced his famous magnetic monopole paper with the following presentiments, which anticipated the future advancement of theoretical physics. He had stressed that the upcoming changes in the style of scientific research promised to be so radical that they may lie outside the shabby boundaries of human intelligence. And the latter will be able to grasp the content of these breakthrough ideas only with the help of the rickety crutches of mathematical speculation. In light of the upcoming shocks, the practical Dirac discreetly proposed a roundabout way, in which 'the most effective method of moving forward' should be

> "to perfect advance the mathematical formalism that forms the existing basis of theoretical physics, and *after* each success in this direction, to try to interpret the new mathematical features in terms

of physical entities (by a process like *Eddington's Principle of Identification*" (Dirac 1931, p.60; my italics).

So, what is this alluring Principle?

The first time the important tenet appears on p.119 of "*Mathematical Theory of Relativity*":

> "Appeal is now made to a *Principle of Identification.* One deductive theory starts with the interval (introduced by the fundamental axiom of &1), from which the [metric] tensor is immediately obtained. By pure mathematics, we derive other tensors…These constitute our world-building material, and the deductive theory aims to construct from this a world that functions in the same way as the known physical world. If we succeed, mass, momentum, stress, etc. must be the vulgar names for certain analytical quantities in the deductive theory; and it is this stage of naming the analytical tensors which are reached in (54.3). If the theory provides a tensor…, which behaves in the same way as the tensor summarizing the mass, momentum, and stress of matter is observed to behave, it is difficult to see how anything could be required of it" (Eddington 1920, p.119).

Moreover, to elicit the content of the abovementioned Principle, Eddington adds:

> "The transition from the geometrical description to the physical description can only be made by identifying the tensors which measure physical quantities with tensors occurring in the pure geometry, and we must proceed by inquiring first what experimental properties the physical tensor possesses, and then seeking a geometrical tensor which possesses these properties by mathematical identities" (Eddington 1920, p.222).

In sum, Eddington refers primarily to the mathematical identification of the vector and tensor structures of his abstruse 'infinitesimal world geometry' with the subtle structures of gravity and electromagnetism (i.e. with the structures of unified theories) pioneered by Hermann Weyl (1919).

The Principle of Identification, as Eddington himself made clear, was *directly inspired by Weyl's* mathematical identification of the vector and tensor structures of his purely infinitesimal world geometry with those of gravitation and electromagnetism. Incidentally, on the 6th of April 1918, even after the GTR, Einstein characterized Weyl's theory as a 'stroke of genius of the highest magnitude.'

The deplorable fact was that eventually the notorious Weyl's 'unified field theory' turned out to be unrealizable. Though this did not prevent Einstein from noting that "the essential point of the theories of Weyl and Eddington" was not that they are geometrical unification but that "they have shown a possible way to represent gravitation and electromagnetism under a unified point of view" (Ryckman 2005, p.238).

Note that famous Eddington's deductive presentation of 'world building' and the accompanying transcendental idealism had been absent from Eddington's first work on relativity, the "*Report on the Relativity Theory of Gravitation*" (1917). Rather, the stimulus rendered by the abovementioned 'mathematical researches' concerns his comprehension of gravity and electromagnetism unification in the context of a 'world geometry.' And the latter conception Eddington discovered in Weyl only *after* the first edition of the report was published (Ryckman 2005, p.182).

It should be added that Arthur Eddington, the Plumian Professor of Astronomy at Cambridge since 1914, and Paul Dirac, a Cambridge post-graduate in physics, were neither trained as philosophers nor widely read in philosophy.

Though one should not ignore the significant circumstance that Dirac had attended lectures of the defiantly *idealistic*, almost platonic philosopher Charlie Dunbar Broad on the philosophical aspects of relativity theory during his study years at Bristol University in 1920/1921. Moreover, direct indications of Dirac reveal that he was not such an ignoramus in the field of philosophy as he preferred to posit himself in the late years, due to age.

> "My attempts to appreciate philosophy were not *very* successful. I felt then that all the things that philosophers said were rather indefinite, and concluded eventually that I did not think philosophy could contribute anything to the advance of physics. *I*

did not immediately have that point of view, but I came to it only *after a lot of thought*, and studying what philosophers said, in particular, Broad" (Ryckman 2005, p.260; my italics).

In sharp contrast to both Dirac and Eddington, Weyl enthusiastically attended Edmund Husserl's lectures while the latter was a professor of philosophy at Gottingen during Weyl's student days there. Moreover, it was Weyl's wife Helene ('Hella') who later came to Gottingen to study with Husserl, who seems to have sparked Weyl's interest in phenomenology.

(Note that Hella Weyl would later be the translator of the inextricable writings of Arthur Eddington and James Jeans into German).

Hence to get a more comprehensive understanding of Eddington's (and consequently Dirac's) interpretation of the GTR one has to dwell into Weyl's distinctive writings. The epistemological ground of these writings is famously constituted by the so-called '*transcendental idealism*' - a cluster of viewpoints commonly associated with Ernst Cassirer, Hermann Weyl, and Arthur Eddington. Its core assertion consists in the tenet according to which 'transcendental constitution of objectivity' in fundamental physical theory, according to a 'transcendental postulate,' in broad generality asserts that "Nature is not thinkable apart from the coexistent subject capable of experiencing that Nature."

On the abovementioned standpoint, Weyl had meticulously reconstructed the objective world of relativity physics within a "purely infinitesimal geometry," corresponding to the phenomenological epistemology wherein only such linear relations as could be present to an infinitesimally bounded spatio-temporal intuition were immediately evident. Within such a differential geometry, Weyl had consequently recasted Einstein's GTR together with Maxwell's theory of electromagnetism. It took place on the privileged epistemological basis of fundamental differential geometric notions having immediate validity only in the tangent space attached to each manifold point P, corresponding to a localized space of intuition. Weyl's distinctive 'theory of gravitation and electromagnetism' is saturated by the phenomenological method of 'eidetic analysis.' In particular, the basic notion of the 'infinitesimal congruent displacement' of a vector has nothing to do with the *actual* behavior of rods and clocks.

According to Husserl's 'the principle of all principles,' in addition to sensory givenness, phenomenology vigorously extends experience, in the form of the 'given,' to intuition, i.e. to nonsensuous, to 'experience in fantasy' or in imagination.

> "That every intuition originally giving (something) should be a validating source of cognition, that everything offered in 'intuition' to us originally (so to speak, 'in the flesh') should be simply taken for that as which it presents itself, but also only within the bounds in which it presents itself here."

Hence phenomenology was aptly characterized by Edmund Husserl as nothing else than "a doctrine of the essences [Wesenslehre] within pure intuition [reiner Intuition]," where one carries out the acts of immediate essential seeing [unmittelbare Wesenschauungen] on the exemplary givenness of transcendentally pure consciousness, and fixes them conceptually.

In apparent opposition to the scientific realism of his day, and in distinctive fashion merging Husserlian 'essential analysis' of space and time as 'forms of intuition' with mathematical constructionism, Weyl sought in this way to proffer a transcendental-phenomenological account of the constitution of the sense of the objective realm of the GTR, the sense of a 'being for consciousness.' Needless to say that it was Husserl's transcendental-phenomenological idealism that was concerned to reveal the "origins of objectivity in transcendental subjectivity."

Hence, in full accordance with Husserl's "*Ideen*" (1913), Weyl's striving to attain a profound philosophical comprehension of the GTR begins with the following assertion:

> "Expressed as a general principle, this means that the real world, and every one of its constituents, with their accompanying characteristics, are, and can only be given as, *intentional objects of acts of consciousness*. The immediate data which I receive are the experiences of consciousness in just the form in which I receive them. They are not composed of the mere stuff of

> perception, as many Positivists assert, but we may say that in a sensation an object, for example, is physically present for me – to whom that sensation relates – in a manner known to everyone, yet, since it is characteristic, it cannot be described more fully. Following Brentano, I shall call it the '*intentional object*'...the intentional object is *immanent*, i.e. like the act itself, it is a real component of my stream of experiences, whereas in the primary act of perception the object is *transcendental*, i.e. it is given in an experience of consciousness, but is not a real component of it. What is immanent is *absolute*, i.e. it is exactly what it is in the form I have it, and I can reduce it, its essence, to the axiomatic by acts of reflection.
>
> On the other hand, transcendental objects have only a *phenomenal* existence; they are appearances presenting themselves in manifold ways and in manifold 'gradations' [...]
>
> It is the nature of a real thing to be inexhaustible in content; we can get an ever-deeper insight into this content by the continual addition of new experiences, partly in apparent contradiction, by bringing them into harmony with one another. In this interpretation, things of the real world are approximate ideas. From this arises the empirical character of all our knowledge of reality" (Weyl [1919], 1922, pp.4-5).
>
> Compare this with Husserl's famous dictum: "My transcendental method ...is the ultimate fulfillment of ...English empiricist philosophy ...to investigate ...the origins of objectivity in transcendental subjectivity..." (Ryckman 2005, p.120).

Accordingly, Einstein's famous 'postulate of general relativity' that freedom to make arbitrary continuous transformations of the space-time coordinates "takes away from space and time the last remnants of physical objectivity" was philosophically grounded by Weil in purely Husserlian terms.

> "In the realm of physics, it is perhaps only the theory of relativity which has made it quite clear that the two essences, space and time, entering into our intuition have no place in the world

> constructed by mathematical physics [...] Just as time is the form of the stream of consciousness, so one may justifiably assert that space is the form of external material reality" (Ryckman 2005, pp.3, 5).

Hence Weyl's "world geometrical" framework for the new physics turned out to be the result of a second-order 'eidetic analysis,' carried out by phenomenological reflection upon the differential geometrical framework of the GTR. Note that the history of Weyl's 1918 unification of gravity and electromagnetism comprises an essential part of the 'context of discovery' of the celebrated 'gauge principle,' one of the most fruitful ideas of XX th century physics.

Eddington, in turn, appeared to be heretical enough to accept Weyl's generalization of Einstein's GTR and to generalize it further, for philosophical reasons essentially similar to Weyl's. As correspondence with Weyl reveals, Eddington turned out to be an early and enthusiastic convert. Probably a significant factor appeared to be that Eddington was born into a Quaker family, and his faith played an important role throughout his life (Kilmister 1994, p.11).

Accordingly, he sought the same goal of constituting the "real world of physics" by reconstructing the GTR within a differential geometry capable of rendering only objects that are a 'synthesis of all aspects' present to all conceivable observers.

Initially, in a letter to Hermann Weyl toward the end of summer 1921, Albert Einstein pronounced on Eddington's unification effort the dim verdict that it was 'beautiful but physically meaningless.' However, later, from aboard ship 'near Singapore' en route from Japan on 11 January 1923, Albert Einstein enthusiastically reported to Niels Bohr:

> "I believe that I have finally understood the connection between electricity and gravitation. Eddington has come closer to the truth than Weyl." (Ryckman 2005, p.220).

Eventually, by 1923 Einstein became one of the enthusiasts of an Eddington-inspired "theory of the affine field," publishing four papers on the subject in that year.

For Eddington, the application of tensor calculus was mandatory since tensors satisfy the relativistic requirement of transformation between all systems of coordinates. Yet the 1928 Dirac equation inspired him to advance his mathematical apparatus. From Dirac's relativistic-electron paper on, Eddington developed what he called 'wave-tensor calculus,' grounded on the Clifford algebra of his E-numbers, an elaborate structure of anti-commuting matrices, and geometries derived from it.

Note that in "*The Nature of the Physical World*," first published in November 1928, Eddington comments on Dirac's philosophical standpoint in connection with the qp-pq=ih/2π equation in the following way:

> "For Dirac, q is a symbol without any kind of numerical interpretation; he calls it a q-number, which is a way of saying that it is not a number at all. I venture to think that there is an idea implied in Dirac's treatment that has great philosophical significance, independently of any question of success in this particular application. The idea is that in digging deeper and deeper into that which lies at the base of physical phenomena we must be prepared to come to entities which, like many things in our conscious experience, are not measurable by numbers in any way; and further, it suggests how exact science, that is to say, the science of phenomena correlated to measure-numbers, can be founded on such a basis.
>
> One of the greatest changes in physics between the 19th century and the present day has been the change in our ideal of scientific explanation. The Victorian physicist boasted that he would not claim to understand a thing until he could make a model of it; and by a model he meant something constructed of levers, geared wheels, squirts [...]
>
> Nowadays ...we turn to the mathematician to built it [the world] *out of his material*. Doubtless, the mathematician is a loftier being than the engineer, but perhaps even he ought not to be entrusted with the Creation unreservedly. We are dealing in physics with a symbolic world, and we can scarcely avoid employing the

mathematician who is a professional wielder of symbols, but we must rise to the full opportunities of the responsible task entrusted to him and not indulge too freely in his own bias for symbols with an arithmetical interpretation. If we are to discern controlling laws of Nature not dictated by the mind it would seem necessary to escape as far as possible from the cut-and-dried framework into which the mind is so ready to force everything that it experiences.

I think that in principle Dirac's method asserts *this* kind of emancipation. He starts with basal entities inexpressible by numbers or number systems and his basal laws are symbolic expressions unconnected with arithmetical operations. The fascinating point is that as the development proceeds actual numbers are exuded from the symbols" (Eddington 1928, p.105).

Anyway, despite all the ups and downs of the Weyl-Eddington unification efforts, it did not obscure the attractiveness of the main source of Eddington's and Weyl's inspiration - the alluring history of the GTR genesis.

Indeed, Einstein's so-called 'geometrization' of the gravitational field in 1914-1915 (the 1913 contrivance of the notorious '*Entwurf*' first together with Marcel Grossman, and then the victorious 1915 GTR) had provided a model (a 'paradigm') of all the entire (geometric) unification of modern physics. The deft contrivance of the GTR basic theoretical scheme was due to the masterful *synthesis* of the 'non-metric' partial theoretical schemes of Abraham, Nordström, and Einstein himself, advanced in 1907-1914 (Nugayev 2018). The staggering amalgamation, which culminated in finding out the stupendous Einstein-Hilbert equations, turned out to be possible only by constructing a basic 'hybrid' or, in A. Eddington's terms, a 'dual' theoretical abstract object - the metric tensor $g_{\mu\nu}$. It is the latter that aptly combines 'geometry' and 'physics.'

Indeed, on the one hand, this tensor is a substantially geometric quantity that masterly characterizes the measured lengths, distances, and times. But, on the other hand, $g_{\mu\nu}$ constitutes ten independent quantities representing the distinctive 'potentials of the gravitational field,' the values of which at each point in space-time are determined by the peculiarities of the distribution of masses and energies, represented by the energy-momentum tensor $T_{\mu\nu}$. Accordingly, the strengths of the gravitational field are described by the curvature tensor $R_{\mu\nu}$, symmetrically containing the derivatives concerning all the 4 coordinates of the metric tensor.

As a result, the hybrid character of the metric tensor as a terse compendium of field quantities is expressed in the fact that it was 'geometrized' in the GTR, i.e. masterfully embedded into the space-time geometry. Herewith, Eddington singled out the important asset of differential geometry in the Weyl-Husserl construction (world building) of the 'physical world': those theoretical objects are 'a synthesis of all aspects about all kinds of observers.' In the popular science monograph '*Space, Time and Gravity*,' which was published in Cambridge in 1920 and anticipated '*The Mathematical Theory of Relativity*' (1923), Eddington underscored:

> "*Physical reality is the synthesis of all physical aspects of Nature*…There may be some illusion in these experimental deductions [i.e. the GTR queer consequences like Mercury perihelion precession or the light bending in the gravitational field of the Sun]; but if not, it must be admitted that the physical reality corresponding to light must be some *synthesis* comprehending both these appearances… *Reality is only obtained when all conceivable points of view have been combined*" (Eddington 1920, p.182; my italics).

Consequently,

> "Ultimately, if this 'synthesis' is done correctly, physical quantities must be represented by tensors in 4-dimensional space-time" (quoted from Ryckman 2005, p.198).

Therefore

> "the four-dimensional geometric *representation* of the world of Minkowski is a 'synthesis' required for taking into account the measurements of observers in all states of uniform motion concerning one another" (Eddington as quoted from Ryckman 2005, p.185).

(VI) And the last operative heuristical principle is epitomized by Weyl's unified theory of electricity and magnetism which was published in 1918 and met first widespread interest. (It was further advanced by Wolfgang Pauli. and taken over into atomic theory by Erwin Schrodinger). Dirac was indirectly influenced by the Weyl-Husserl epistemology not only via Eddington's numerous writings but through his first-hand acquaintance with Weyl's unified theory of electricity and gravitation.

One should remember that Einstein's so-called 'geometrization of gravitational interactions' in 1915 proffered the modern geometrical unification program in physics and that the program successfully started in 1918 with Weyl's startling broadening of the geometry of the GTR into a 'theory of electricity and gravitation.' Hence it is no wonder that Dirac was 'just fascinated by it.' His genuine interest in Weyl's theory stemmed from his post-graduate days in Cambridge when he had studied it thoroughly. By the wide-spread opinion, he had concluded that this theory, "although mathematically appealing, was physically unsound." Note that he knew German well enough to edit the translation of his "*Principles of Quantum Mechanics*" into German in 1931. Nevertheless, Dirac continued to be fascinated by the fine mathematical structure of the theory, as well as by Weyl's general philosophy of physics, which he appreciated as representative of his ideal of a "powerful method of advance." The true aesthetic ground for Dirac's high praise of Weyl's unified theory was made clear in 1973 when he returned to cosmology. Dirac overtly characterized the Weyl theory as "a very ***beautiful synthesis*** of the electromagnetic field and the gravitational field," a theory that "remains as the outstanding one, unrivaled by its *simplicity and beauty*" (Dirac 1973a, p.53).

In complete agreement with the abovementioned heuristic principles, Dirac stressed:

> "It appears as one of the fundamental principles of Nature that the equations expressing basic laws should be invariant under the widest possible group of transformations… The passage to *Weyl's geometry* is a further step in the direction of widening the group of transformations underlying physical laws. One now has to

consider transformations of gauge as well as transformations of linear coordinates and one has to take one's physical laws to be invariant under all these transformations, which impose stringent conditions on them" (Dirac 1973b, p.418).

One should add that, while visiting Germany, Dirac went to a course of lectures on group theory given by Hermann Weyl in February 1927. The interrelations between Dirac and Weyl and the level of admiration expressed by Dirac to Weyl's ideas can be illustrated as follows. As is well-known, after the publication of his 1928 masterpiece on the relativistic electron, Dirac had encountered with a (±) difficulty connected with the interpretation of negative solutions of the equation $E= \pm c\sqrt{p^2 + m^2 c^2}$.It is important that Dirac's idea to answer the difficulty was indebted to the suggestion made by Weyl in the spring of 1929.The suggestion was related to a unitary theory of gravitation, electromagnetism, and quanta in which the electromagnetism was brought in as an accompaniment of the quantum wave field and not, as in other unitary theories, of the gravitational field.

Weyl's sober proposal was quickly taken up by Dirac in his paper "*A Theory of Electron and Proton.*" Moreover, in the second edition of his epoch-making "*Gruppentheorie und Quantenmechanik*" Weyl had revealed that according to Dirac's theory of the electron the hole must necessarily possess the same mass as an ordinary electron. When, in 1931, Dirac gave up his unitary theory and proposed the anti-electron as a separate article, he quoted Weyl's suggestion as a decisive argument. It is not an accident that in his letter to Igor Tamm, 3 January 1929, Dirac noted that Weyl's '*Gruppentheorie und Quantenmechanic*' (1928) is

"very clearly written and [...] far the most connected account of quantum mechanics that has yet appeared" (Kragh 1990, p.79).

(VII) Yet it's small wonder that the most controversial trait of Dirac's heuristic is his '***profound intuition***' commonly contrasted to 'logical clarity in the presentation.' It was always Dirac's ideal, in formulating a physical theory, to be able to construct or to discover a 'reasonable mathematical scheme' and to interpret the equations at hand most directly and naturally. Of course, "the scheme must be mathematically well-defined and self-consistent," and in agreement with well-established principles; yet these criteria are quite insufficient even for choosing a definite scheme among many. All appearance, to comprehend the problem situation, one should introduce something like Husserl's 'principle of all principles,' extending, in addition to sensory givenness, experience, in the form of the 'given,' to intuition, i. e. to nonsensuous, to 'experience in fantasy' or in imagination.

> "That every intuition originally giving (something) should be a validating source of cognition, that everything offered in 'intuition' to us originally (so to speak, 'in the flesh') should be simply taken for that as which it presents itself, but also only within the bounds in which it presents itself here."

Hence too many of Dirac's great works were based on ideas that came unexpectedly for him – from so to speak 'out of the blue' (Kragh). Needless to say that Dirac himself could not afford the liberty of direct indicating the importance of his intuition in physical research. Others did it for him. In his review of "*The Principles of the Quantum Mechanics*" the mathematician Bernard Osgood Koopman found that these principles were more characterized by Dirac's *profound intuition* than any logical clarity in presentation. (Kragh 1990, p.79).

Likewise, the American physicist and philosopher of science Henry Margenau noted that

> "While Dirac presents his reasoning with admirable simplicity and allows himself to be guided at every step by *physical intuition* – refusing at several places to be burdened by the impediment of mathematical rigor" (Kragh 1990, p.79).

According to Pauli's sarcastic 1935 remark,

> "The success seems to have been on the side of Dirac rather than on logic" (Kragh 1990, p.112).

Further, at a conference in honor of Dirac in 1983, one of his pupils (Harish-Chandra) observed that his former supervisor always preferred to rely on his intuition rather than on established knowledge (Kragh 1990, p.251).

V.3.The Substantial Milestones of Successful Mature Theory Genesis.

Let me recap those main stages of the genesis and becoming of any successful mature theory that played a decisive role in the creation of Dirac's masterpiece as well.

(i) The inevitable 'encounter' of stiff and diverse pivotal 'mathematical projections of nature' (such as statistical mechanics vs. Maxwellian electrodynamics; or Newtonian theory of gravitation vs. STR, or STR vs. nonrelativistic QM) and subsequent step-by-step contrivance of a medley of 'hybrid' theoretical schemes. A hybrid theoretical scheme possesses advantages of its own. It describes its subject area so that there is a pressing need for their synthesis and for constructing a global theoretical scheme with the relations between the theoretical objects related by the global theory basic equations. In the GTR genesis paradigmatic case, these hybrid theoretical schemes were represented by the artful non-metric theories of Abraham, Nordström, and Einstein (Nugayev 2020). In the case under consideration – the Dirac synthesis - the hybrid schemes were incarnated by Klein-Gordon-Fock (1926) and Pauli-Heisenberg-Jordan (1926) half-phenomenological theories.

(ii) It commonly turns out that the blunt resolution of the problem situation via elevating the basic objects of any one of the constructed hybrid theoretical schemes on the pedestal of the global theory simply 'does not work.' (Commonly due to explanatory problems of the global reductionist theory in embracing significantly heterogeneous domains of validity). Eventually, the understanding comes that the desired synthesis is possible only via the 'radical transformation' (Dirac), i.e. due to the involvement of some substantially novel branch of mathematics. Its conceptual system should be able to generate a 'neutral language' vocabulary for describing, comparing, and 'reconciling' all the hybrid schemes involved (Cassirer 1910).

For instance, when Maxwell had hardly contrived his celebrated *fundamental theoretical scheme* (Stepin 2006) of classical electrodynamics, the subtle language of vector calculus (embodied in the equations of hydrodynamics with its operations div **E** and rot **H)** started to function as a genuine neutral language (Nugayev 2015). Similarly, in the GTR genesis, the subtle notions of Riemannian geometry and tensor calculus also commenced to function as an efficacious neutral language; withal tensors constitute the geometric objects, the relations between which are preserved in the transition from one non-inertial frame of reference to another.

In the Dirac theory genesis, the role of such a mellow language was played by the odd language of spinor calculus with its Clifford algebra, primarily referred to in the works of W. Pauli in 1926.

(iii) The neutral language obtained is used to generalize all the available hybrid schemes and arrive at the simplest equation embracing all the hybrid regularities encountered (Einstein &Grossman's '*Entwurf*'; Dirac's relativistic electron equation). Such a generalization commonly takes place in one of the encountered hybrid theoretical schemes, and has a *local* character at first, but eventually acquires a *global* significance. The object lesson is Abraham's deft conjecture, which generalized the four-dimensional Minkowski metric to the case of the metric tensor dependence on coordinates $ds^2 = dx^2 + dy^2 + dz^2 - c^2dt^2$, or the introduction of the four-dimensional function $\psi = \psi\ (x_0, x_1, x_2, x_3)$ by Dirac.

(iv) The decisive stage is taking place. Among the mathematical objects, the relations between which are described by the proposed equations of the global theory, one should discover an object that possesses the properties of the basic object of the global theoretical scheme. Viz., it is necessary to reveal the 'dual' (Eddington) or 'hybrid' (Nugayev) nature of the theoretical object. Viz., to demonstrate that it simultaneously possesses certain 'geometric' properties, transforming itself as a tensor $g_{\mu\nu}$ and, on the other hand, has 'physical' properties also, since it is transformed as an irreducible representation of the Poincaré group, the Lorentz group, etc. This elucidation of the so-called 'sheer physical,' operational sense of mathematical terms should allow comparing the obtained quantitative relationships with experimental data.

Accordingly, in Dirac's relativistic electron theory, the wave functions are described by bispinors ψ, the components of which are complex numbers. Moreover, both in Dirac's theory, as well as in QFT in general, bispinors are connected with such fundamental particles as electrons and even quarks. Under transformations of the Lorentz group, which describe the symmetries of the Minkowski space, the bispinors are transformed as specific irreducible representations of this group - as the set of operators in the Hilbert space. It is the symmetries of flat space-time that famously allow, by Noether's theorem, electrons, and quarks to possess such conserved properties as mass, spin, and charge.

Consequently, ψ - the pivotal mathematical construct of the Dirac equation - has an explicit *dual structure*. On the one hand, it is a typical bispinor. As such, when the coordinate system changes (including rotations in 3-dimensional space and transformations between frames of reference moving rapidly relative to each other), it is transformed not as a vector of ordinary space, but as a mathematical object in a special space called 'internal,' not intersecting with the usual, 'external' space.

But, on the other hand, the same ψ represents, at least in the first approximation, the amplitude of the probability of one electron, associated with the density of the charge and current of the Dirac particle. By the conservation of charge, the quantity that is taken to be the total probability of finding a particle is also conserved. Accordingly, Dirac's theory not only featured the geometric properties of bispinors, but was also able to predict the magnetic moment of an electron, as well as the fine structure of lines in the spectrum of atoms, and, famously, the existence of antiparticles (or positrons).

V.4. The Milestones of Dirac's Startling Electron Theory Contrivance.

By 1928, business in the inextricable realm of quantum theory proceeded as follows. The primary pioneering attempts by Planck (1900), Einstein (1905), and Bohr (1913), related to the so-called 'old quantum theory,' gave way to a relatively calm stage of accumulation of abundant experimental data within the framework of the new paradigm. The stage culminated in the contrivance of two perfect and complete NQM versions proffered by Schrödinger and Heisenberg in 1926. Dirac (1927) had famously laid out that these options are empirically equivalent. For obvious reasons, the subject was broached on the creation of relativistic QM. Namely, any correct quantum theory should satisfy the requirement of relativity: the laws of motion that take place in one inertial frame of reference must also be fulfilled in all the other inertial ones. Relativistic quantum theory must be formulated in a Lorentz-invariant form (Bjorken & Drell 1998).

> "This is necessary in order that the theory may apply to high-speed particles" (Dirac [1930], 1981, p.253).

As a result of the first, blunt attempts to join the heterogeneous principles of SRT and NQM, the following hybrid theoretical schemes were contrived. First, Erwin Schrödinger himself, along with deriving his renowned equation in the spring of 1926, attempted to obtain a kind of its relativistic prototype by inserting relativistic corrections to the Hamilton-Jacobi equation for an electron in an electromagnetic field. Indeed, STR requires that the energy of a free particle be described by the expression $H^2 = c^2\boldsymbol{p}^2 + m^2 c^4$.

To get a QM analogue of the equation, one must replace the corresponding variables with operators representing the derivatives in all 4 coordinates. So, one has

$$\Delta \psi - (1/c^2)\, \partial^2\psi/ \partial t^2 - m^4c^4/h^2 = 0 \qquad (1)$$

This relation, also derived by several scholars, back in 1926, was christened the 'Klein-Gordon-Fock (KGF) equation.' In the textbooks, it is taken as an equation describing the motion of a scalar particle, and the failure of the KGF equation is commonly associated with its inability to theoretically reproduce the particle spin. But that is inaccurate. The KGF equation should be comprehended from the very beginning as the equation of a *spinless* particle. Its abandonment by Schrödinger was, however, because the equation could not exhibit the strictly one-particle system: it did not preserve what non-relativistic QM considered a probability of particle occurrence anywhere in space. Yet later this property was attributed to the fact that relativistic quantum theory does not preserve the number of particles. It takes the particle as a quantum of fields that can arise and wear off while maintaining the total energy. Although initially Dirac's equation seemed appropriate to portray a system of a single electron (it preserves probability), today one is aware that it features quantum electron-positron field (similar to the KGF equation that describes the field of scalar particles).

On the other hand, in 1926 Wolfgang Pauli aspired to bolster the spin hypothesis put forward by Uhlenbeck and Goudsmit which qualitatively adequately explained the spectral data related to the so-called 'anomalous Zeeman effect.' For this he aptly modified the Schrödinger equation by introducing into the Hamiltonian the terms containing the so-called 'spin operators' described by 2-dimensional matrices (Pauli 1926). In a nutshell, Pauli's sober methodology can be summarized as follows. Instead of the unperturbed Hamiltonian H_0, representing the basic state of the electron as a quantum system, the final Hamiltonian $H = H_0 + H_1 + H_2$ was contrived. Here H_1 exhibits the contribution from the magnetic field and H_2 is the contribution characterizing the electron spin. Alas, despite the empirical success, in theoretical respect this theory, even masterly amended by Heisenberg & Jordan (1926), turned out to be deficient also.

As a result, by 1928, the problem situation was as follows. Two genuine hybrid theoretical schemes had been tentatively proposed that strived to merge SRT and NQM by introducing stale relativistic ad hoc corrections to certain parts of the Hamiltonian to arrive at the Schrödinger equation exhibiting a relativistic electron. But all these attempts failed. On the one hand, the partial theoretical KGF scheme, perfectly consistent with the STR principles, diverged not only from experimental data but – what's worse - from the NQM empirically sound principles.

On the other hand, the Pauli-Heisenberg-Jordan abstruse theoretical scheme, which perfectly explained the splitting of spectral lines and, therefore, was consistent with the NQM bizarre principles, diverged, in turn, with the fine principles of STR.

The general reason for the failure was the hybrid, temporary nature of the results obtained. The latter were worked out through merely *partial* modifications of entrenched theoretical traditions. An embracing and powerful approach was needed, i.e. the initiation of some general heuristic principle, analogous to Einstein's efficacious 'principle of equivalence' (with the identity of gravity and inertia), and its implementation in a sequence of more and more stringent ideal models. The consequent application of the principle was to gradually lead to the discovering of general equations describing the relationship between the basic mathematical constructs. Raising these constructs to the rank of abstract objects of the global theoretical scheme, and their subsequent operationalization should have made it possible to derive such novel consequences that could be compared with experimental data. All the indispensable requirements were met by Paul Dirac's breakthrough '*The Quantum Theory of the Electron.*' The paper was received by the editorial staff of the '*Proceedings of the Royal Society*' on January 2, 1928.

In Dirac's distinctive theory, the role of the 'ontological' (in the GTR case) 'principle of equivalence' was played now by the epistemological 'principle of mathematical beauty.' The principle in many respects was reduced, as mentioned earlier, to the requirement of making *minimal*, in comparison with classical theories, changes in a new, non-classical theory. But this semantics cannot exhaust, in my view, the content of this heuristic, most important for Dirac, principle. It is also necessary to take into account its 'synthetic,' Eddingtonian and Weylean, subtle facets, connected with the '*beautiful synthesis*' (according to Dirac's appraisal of Weyl's unification of gravity and electromagnetism). The latter consists in the applying, when formulating one's equations, such a mathematical language that is capable of realizing a true synthesis of 'geometry' and 'physics.' The unification is achieved by contriving mathematical constructs that effectively *integrate* both mathematical, 'transformational,' and purely physical, 'material' properties.

Whereupon, of certain interest is the following hypothesis of the prominent 'Dirac scholar' - the Danish historian of science Helge Kragh. While constructing the basic theoretical scheme of his theory, Dirac may have turned to the work of German mathematicians of the XIXth century related to Clifford algebras (Kragh 1990). Another source of mathematical inspiration (Mehra & Rechenberg 2000, p.294) could have been for Dirac the ideas of the English mathematician Henry Baker, outlined in his '*Principles of Geometry*,' to which Dirac referred back in 1925 during the advancement of his q-number version of QM. In the book, Dirac could discover the 2x2 matrices (similar to Pauli matrices), and even the 4x4 matrices similar to those properly used in the 1928 paper.

Be that as it may, the 'mathematical beauty' consists not only in providing the maximum number of results with minimal means but also in the ability to *embrace* the ideas that at first glance seem to be in no way related to each other with single elegant gesture.

Let us recall how Dirac overtly commended the Weyl theory as "a very *beautiful synthesis* of the electromagnetic field and the gravitational field," a theory that "remains as the outstanding one, ***unrivaled*** by its *simplicity and beauty*" (Dirac 1973a, p.53).

Needless to recall that the need for synthesis played a special role for Dirac:

> "I find this position [the paradox in the mechanical worldview caused by the peculiar role of initial conditions] very unsatisfactory *philosophically*, as it goes against all ideas of the *unity of nature*" (Dirac 1938-39, p.127; my italics).

Examples of such an integral mathematical language are Newton's and Leibnitz's differential and integral calculus, the vector calculus of hydraulic Maxwellian models, as well as the GTR tensor calculus used by A. Einstein. It is hard not to recall Dirac, who expressed himself in his single genuinely philosophical article '*The relationship between mathematics and physics*' clearly and frankly as usual:

> "Pure mathematics and physics are becoming more closely connected, though their methods remain different. One may describe the situation by saying that the mathematician plays a game in which he invents the rules while the physicist plays a game in which the rules are provided by Nature, but as time goes on it becomes increasingly evident that the rules which the mathematician finds interesting are the same as those which Nature has chosen" (Dirac 1938-39, p.127).

Pauli's artful theory could have facilely become the starting point for the construction of the relativistic theory of the electron: it was merely necessary to formulate it in a Lorentz-invariant way, generalizing the 2x2 matrices to the 4x4 case. But Dirac claimed more.

The first stage of his research consisted in the introduction of a substantially novel mathematical construct: a 4-component wave function $\psi = (\psi_1, \psi_2, \psi_3, \psi_4)$. This move was extremely bold and risky since empirical indication in favor of such a generalization was absent completely. The spin of an electron could only take two values, not four.

The next steps, of course, consisted in both relativistic generalizations of two-dimensional Pauli matrices to 4 dimensions, and in the symmetric ranking of all derivatives, which should have looked, by relativistic invariance, in the same way, etc. (The technical details are proffered by related textbooks like Bjorken & Drell, 1998, or better yet, by meticulous historical-scientific accounts like Kragh 1981 and Valente 2020). As a result, Dirac arrived at the following relativistic equation for a free electron

$$[p_0 - \rho_1 (\boldsymbol{\sigma},\mathbf{p}) - \rho_3 mc]\,\psi = 0, \qquad (2)$$

where $p_0 = ih\,\partial/c\partial t$, $\boldsymbol{p} = (p_1,p_2,p_3)$; $p_r = - ih \frac{\partial}{c\partial x^r}$, $r = 1,2,3$; $\boldsymbol{\sigma} = (\sigma_1, \sigma_2, \sigma_3)$.

The exact cumbersome expressions for σ and ρ (4x4) matrices can be found in Dirac's renowned book ([1930], 1981) on pp.253-255.

Further, Dirac generalized this equation in the most common way to the case of an electron in an electromagnetic field via the standard replacement of p_0 with $p_0 + (e/c)\,A_0$ and $\mathbf{p}$ with $\mathbf{p} + (e / c)\,\mathbf{A}$. Having squared it, he obtained a significant expression, which contained, in comparison with the Klein-Gordon-Fock and Pauli equations, two completely new terms. Naturally Dirac identified the first term with the electron magnetic moment $(eh / 2mc)\,\boldsymbol{\sigma}$. As for the second term, which related to the electric moment, Dirac easily and cheerfully declared it physically meaningless as containing an imaginary complex number that cannot be part of the real Hamiltonian.

This seemingly nondescript detail drops an important hint: Dirac's theory is by *no* means the result of any irreproachable conclusion either from an experiment or from a certain deep *a priori* principle. This is a quirk mathematical surmise, the validity of which still needs to be confirmed by the results of future experiments and calculations. Incidentally, in chapter XI "*Relativistic Theory of Electron*" of his prominent "*Principles of Quantum Mechanics*," first published in 1930, Dirac had succinctly reproduced the derivation of his pioneer 1927 paper, humbly admitting that its exact form is 'just' a guess, albeit well educated and well supported (Dirac [1930], 1981, pp.253-75)

In sum, the successful synthesis consisted in contriving a miraculous equation, the space-time properties of which matched the STR, while the quantum properties reconciled with the NQM transformation properties. Surely, the main asset of Dirac's theory consisted in the theoretical reproduction of the electron spin, which turned out now to be a *strictly relativistic consequence* of his theory.

> "[Initially] I was not interested in bringing the spin of the electron into the wave equation… It was a great surprise for me when I, later on, discovered that the simplest possible case did involve the spin" (Dirac quoted from Kragh 1981, p.55).

An unconstrained derivation of the correct spin and magnetic moments of an electron allowed Dirac to naturally explain away all the spectroscopic puzzles at hand - the Zeeman and the Paschen-Buck effects, as well as the Sommerfeld odd formula. Yet, perhaps, the most unexpected consequence of Dirac's distinctive theory was the staggering prediction of a new type of matter - the so-called 'antimatter,' the existence of which was not expected by anyone before, but which was miraculously confirmed several years later. Only *after* this amazing corroboration did all the legitimate doubts about the validity of the Dirac synthesis of STR and QM completely disappear.

Associated with this and subsequent achievements is such a hallmark of the work of the former electrical engineer as 'gradualness' (piecemealness). For instance, to the mathematician Klein, who frivolously tried to unite NQM and GTR in one step, Dirac acrimoniously retorted that the main reason for his failure was that he was trying to solve too many problems at once.

> "One should not try to accomplish too much in one stage. One should separate the difficulties in physics one from another as far as possible, and then dispose of them one by one" (Dirac quoted from Kragh 1981, p.40).

This is exactly what Dirac did. Yet the introduction of spin turned out to be only one of the main benefits of his miraculous equation. The equation naturally ('by itself') explained the spin magnetic moment of an electron that Pauli's strained equations had inserted only phenomenologically. In this sense, Dirac's spellbinding equation constituted the first consistent description of particles with spin $\frac{1}{2}$. Though the most amazing implication of Dirac's equation, which is inherently encoded in using 4-component bispinors, consisted in the prediction of antiparticles. Though initially he had misinterpreted them, eventually he anticipated them correctly. To go from the whimsical relativistic electron equation to the solid and mature theory of antiparticles several non-trivial 'empirically-progressive problemshifts' were needed. It was necessary to move from a description of plane quantum systems with a fixed number of particles to an inextricable 'second quantization' of physical fields. Without this crucial step, the relativistic quantum theory was blatantly inconsistent.

It might be worth mentioning Dirac's major contributions to the contrivance of the basic mathematical apparatus of quantum theory, which gradually appeared in re-editions of his renowned "*The Principles of Quantum Mechanics*." It should be stressed that Dirac with his distinctive *intuitive* approach foreshadowed the sort of mathematics that was rigorously contrived only many years later (specifically, the theory of distributions, as well as the theory of so-called 'rigged Hilbert space').

V.5. The Corollaries and Prospects of Research.

Chapter XI "*Relativistic Theory of the Electron*" of Dirac's masterpiece "*The Principles of Quantum Mechanics*" (1981) first published by the Clarendon Press in 1930, commences as follows:

> "Let us now try to make the theory invariant under Lorentz transformations, so that it conforms to the special theory of relativity. This is necessary so that the theory may apply to high-speed particles" (Dirac [1930], 1981, p.253).

It is clear that this unobtrusive scheme had put up the structure of Dirac's pioneering paper '*The Quantum Theory of the Electron*' (1928, '*Proceedings of the Royal Society*' *(London*). The fact serves as a vivid illustration of the need to distinguish between the '*logic of discovery*' and the '*logic of confirmation*' (the logic of justification) of a scientific hypothesis, incumbent at the heart of the philosophy of science.

Nevertheless, the *actual* 'logic of Dirac's discovery' turns out to be completely different, as it was elicited, for instance, thanks to plain interview given by Dirac to the historian of science Jagdish Mehra much later.

> "I was *not* interested in bringing the spin of the electron into the wave equation, did not consider the question *at all,* and did not make any use of Pauli's work. The reason for this is that *my dominating interest* was to get a relativistic theory agreeing with my *general* physical interpretation and transformation theory. I thought that this problem should first be solved in the simplest possible case, which was presumably the spinless particle, and only *after* that should one go to consider how to bring in spin" (Dirac 1977, quoted from Mehra & Rechenberg 2000, p.290).

Therefore, the hallmark of Dirac's stupendous synthesis was that it turned out to be an unconstrained *continuation* of his previous abundant and profound studies, which consisted, as is well-known, in successful attempts to exhibit the empirical equivalence of the Heisenberg and Schrödinger NQM versions by advancing his ingenious approach. The latter scrutinized (from a unified standpoint!) transformation properties of 'bra' and 'ket' vectors in Hilbert space (Dirac [1930], 1981).

The second decisive trait was the fact that the sought-for synthesis of STR and quantum mechanics was worked out according to the most developed model for that time - reproducing the GTR genesis. In the latter case the efficacious unification of 'geometry' and 'physics' turned out to be possible due to the use of a fundamentally new mathematical apparatus - tensor calculus. The latter played the role of a neutral theoretical language for Einstein, which made it possible to compare and therefore unite the heterogeneous theoretical schemes of non-metric theories of Abraham, Nordström, and Einstein himself. As Dirac himself later stressed in his profound paper '*Quantum singularities in an electromagnetic field*,'

> "The steady progress of physics requires for its theoretical formulation a mathematics that gets continually more advanced. This is only natural and to be expected. What, however, was not expected by the scientific workers of the last century was the particular form that the line of advancement of the mathematics would take, namely, it was expected that the mathematics would get more and more complicated, but would rest permanently of axioms and definitions, while actually, the modern physical developments have required a mathematics that continually shifts its foundations and gets more abstract. Non-euclidean geometry and non-commutative algebra, which were at one time considered to be purely fictions of the mind and pastimes for logical thinkers, have now been found to be very necessary for the description of the general facts of the physical world. It seems likely that this process of increasing abstraction will continue in the future and that *advances in physics* are to be associated with a continual modification of the axioms at the base of mathematics rather than with a logical development of any one mathematical scheme on a fixed foundation" (Dirac 1931, p.60; my italics).

Just like Einstein generalized the available theoretical schemes using the vocabulary of tensor calculus, Dirac generalized the theoretical schemes of Klein-Gordon-Fock and Pauli-Heisenberg-Jordan using no less powerful and advanced language of the theory of bispinors and Clifford algebra.

And, finally, the third hallmark of synthesis consisted in the fact that Dirac ascertained the specifics of the GTR synthesis not by dint of personal contacts with its creator or sedulous scrutinizing of primary sources. He did it through the peculiar prism of his Cambridge colleague and mentor - the English astronomer Sir Arthur Eddington, the author of the best (according to Einstein himself) exposition of STR and GTR - monograph '*Mathematical Theory of Relativity.*' Note that for Eddington the mathematical constructs of geometry – all those tensors, vectors, and 'square roots of vectors' - were the *primordial bricks* for stout "World Building."

> "We are going to build a world – a physical world which will give a shadow performance of the drama enacted in the world of experience" (Eddington [1928], 1948, p.128).

The world-building is grounded on 'synthesis of representations' (in line with Husserlian notable ideas) in the spirit of the quirk theories of Cartan and Weyl. Therefore, having conceived the methods of constructing dual (hybrid) objects, containing, among other things, recipes for 'interpreting new mathematical properties in terms of physical quantities (using a process similar to Eddington's Identification Principle),' Dirac was in no hurry to reveal the underlying philosophical principles of Eddington's philosophy. The odd principles evoked an extremely negative reaction from the majority of Copenhagen-minded physicists; the stout positivist Wolfgang Pauli even christened them - albeit in one of his letters, not publicly! – 'complete nonsense.' Although, when it came to other Eddington influences that had a significant impact on the science of the XX century. - whether it was the staggering cosmological 'hypothesis of large numbers,' which led to the anthropic principle, or the adventurous δ-function - Dirac did not hesitate to reveal Eddington's influence:

> "I noticed the need for it [i.e. the δ-function] very much in a certain place of Eddington's book on '*The Mathematical Theory of Relativity*' [i.e. in a footnote on p.190 of the 1923 edition]" (quoted from Mehra & Rechenberg 2000, p.81).

Therefore, imperatively emphasizing the pivotal role of the 'principle of mathematical beauty' in the construction of new theories, Dirac cautiously noted that the latter is inherent in all the objects of Nature and therefore can be disclosed by the theoretician in his theoretical models. True, the question inevitably arises of why Nature is arranged in this way, and who nevertheless tried to arrange it in this way too. Maybe the answer can be found in the following dictum:

> "one could perhaps describe the situation by saying that God is a mathematician of a very high order, and he used very advanced mathematics in constructing the universe" (Dirac 1963, p.45).

CONCLUSIONS

Briefly, the main results of the monograph can be summarized as follows.

1. Since Alexander Koyrè and Thomas Kuhn, the Ptolemy-Copernicus breakthrough is taken as the 'paradigmatic case of all paradigms,' and many weighty history & philosophy of science problems and their solutions are commonly contemplated precisely on this material. In particular, at least some of the drawbacks of well-known epistemological models of the Copernican revolution turn out to be connected with one and the same reason – lack of proper scrutiny of the *subtle interconnections of cognitive and social aspects* of scientific research. The epistemological models, as a matter of fact, ignore to varying degrees the hard fact that science is a '*community effect*.' On the contrary, according to the romantic 'great genius' standpoint, the startling insights of very peculiar individuals, the 'chosen ones of the gods,' are what drive science forward.

2. It is maintained that this is a distorted picture and the goal of the **first chapter** of the monograph is to strengthen efforts in landing the sky-high epistemological models of scientific revolutions. The first step to answer the question had been taken by Thomas Kuhn's 1962 "*Structure of Scientific Revolutions*" with its path-breaking reconciliation of cognitive and social factors. Yet Kuhn's model turns out too *narrow* with its blunt reduction of social factors to social-psychological ones, - on the one hand, - and notorious 'incommensurability tenet,' - on the other. Kuhn's distinctive interpretation of incommensurability as a matter of fact impugns progress in science.

3. Hence in the wake of Kuhn's '*Structure*,' amended by Michel Foucault's 'reflections on power' and Joseph Rouse's 'practical hermeneutics,' as well as Bruno Latour's sober sociology of knowledge, solid writings of Stephen Shapin, Philip Kitcher and Helen Longino, a further step is taken on the path of explanation for the mundane reasons for the incipience and triumph of the Copernican breakthrough program.

4. Nevertheless, further gradual reconciliation of cognitive and social approaches necessarily demands the expansion of the epistemological base of research. Hence it is contended that more intensive and *conscious* involvement of phenomenological ideas presented by Edmund Husserl and Martin Heidegger is required . The history of human cognition can be taken as a series of epochs of 'unconcealment' with integrity of each one provided by reconciliation, interaction and interpenetration of sundry *practices.* As a result, amalgamation of SSK with congenial components of Heidegger's existential epistemology leads to account of science as 'social existence' (Kochan's term). This amalgamation is grounded on the genuine phenomenological deconstruction of the Cartesian Subject-Object distinction in its entirety. To provide a paradigmatic example, recall how Werner Heisenberg was deeply preoccupied with the "abyss opened by the unsought after discovery of the inseparability of subject and object in quantum theory" (see his 1959 comments in '*Martin Heidegger zum Siebzigsten Geburtstag*'). Furthermore, in a 1949 essay on the '*Relations of Theoretical Physics to Heidegger's thinking*' von Weizsäcker noted that

> "the substantive separation of the *res cogitans* and *res extensa* is the methodological presupposition of the classical natural sciences in their entirety" (von Weizsäcker [1949], 1990, p.244).

Likewise, inextricable interaction of social and cognitive factors in the advancement of science may be described in the quantum-mechanical terms. In the revolutionary stages there are several cognitively admissible ways of the development of science (several ground states of the Ψ – function). Yet the social-cultural factors (the act of measurement) choose resolutely only one option. The other vivid example of harmonic balance of social and cognitive factors is a theory-choice situation when the influence of social-cultural factors changes the 'weights of cognitive ones'

(Kuhn 1957.Anyway, comprehension of social existence of science needs further investigation).

5. As a result, it is elicited that current epistemological accounts of the Ptolemy-Copernicus breakthrough miss a substantial point: the alternative ontologies of Ptolemy (mathematical astronomy) and Aristotle (qualitative physics) arose out of diverse social practices. Ptolemy's wayward ontology was grounded in pagan religious meditative practices (the divine mathematics of the Skies), and Aristotle's solid worldview was entrenched in sensual and object-manipulative ones (the mundane physics of the Earth). Correspondingly, the Copernican Revolution is envisaged in the context of intense interaction and interpenetration of Aristotelean and Ptolemaic sophisticated research practices that take root in certain material and spiritual practices of antiquity. Respectively, Copernican and Ptolemaic congenial research programs were realizing the radically different from each other ways of astronomy and physics stout reconciliation.

For many centuries, the dualism had been 'frozen up' in the Aristotle-Ptolemy pagan cosmology. Eventually, the Aristotle – Ptolemy pagan cosmology could not help but be exposed to repeated cogent attacks during the Middle Ages since it apparently confronted the renowned principles of monotheism, not admitting the impervious demarcation line between the celestial and mundane realms. All the opposite worlds should have one and the same Creator. Commencing with the unification, Copernicus, in effect, paved the way for the descent of mathematics from Heaven to Earth and the spread of natural philosophy from Earth to Heaven.

Following Edmund Husserl's studies, Alexandre Koyrè maintained that modern science has succeeded in overcoming the split, inherent in Aristotelian science, between Earth and Space. The Christian canon Nicolaus Copernicus blew the cosmology up from within contending that Earth is a common planet of the solar system. The explosion was performed in a quite different (monotheistic) socio-cultural context, and, as a result, Galileo set up a throng of corresponding research practices aimed at the efficient reconciliation of mundane physics of the Earth with sublime mathematics of the Skies.

6.Thereafter, an attempt is made to *generalize* the lessons of the Ptolemy-Copernicus paradigmatic transition and to propose an outline of a scientific revolution model. According to the outline, profound breakthroughs in science were *not* due to ingenious contrivances of brave novel paradigms, invention of 'send by God' immaculate ideas 'ex nihilo.' The breakthroughs were caused, among other things, by the harrowing piecemeal processes of accommodation, interpenetration, and intertwinement of the 'old' research practices preceding the radical breaks.

7. In the **second chapter** the outline of a scientific revolution is applied to the case of modern mathematized science or 'mathesis.' As a result, a comprehensible ideal model that provides analytical tools for studying the mechanisms by which the cognitive structures of scientific knowledge evolve and interact with each other and with social dimensions of the historical advancement of science is posited. According to the model, an origin of a scientific revolution consists not in well-trodden discrepancies between bold theoretical speculations and stubborn 'hard facts,' but in deadly contradictions of mature theories from 'old' rival scientific research programs with each other. In this case the origin of scientific revolution is constituted by the clash of different '*mathematical projections of nature*' with each projection represented by distinctive bundle of theoretical and empirical practices organized into a single whole by the basic idealizations model (or the so-called 'Fundamental Theoretical Scheme' in Stepin's vocabulary). In the case of modern science a 'mathematical projection of nature' is typified by a mature scientific theory with its systems of theoretical and empirical schemes described by Vyacheslav Stepin.

Hence a mature scientific theory constitutes a *hierarchy* of practices disposed at three levels. The mature theory structure consists of fundamental and partial theoretical schemes, as well as the level of empirical schemes with the directions of advancement governed by common 'basic blueprint,' Grundriss of the project. Within the project, theoretical and experimental practices come to mutually reinforce one another, each giving strength to the other. The influence is 'reciprocal and dialectical.'

8. It is disclosed that a mature theory can be 'refuted' only when it is impelled to explain a 'critical experiment' on a par with *another* different and contradictory mature theory. The inextricable problem situation arises from the encounter of two mature theories. The latter, in the inexorable process of their expansion, meet the experiment that belongs to the domains of validity of both theories. An apparent example is the renowned experiments related to blackbody radiation that demanded for their theoretical reproduction and explanation the joint application of dainty Maxwell's electrodynamics, daring Boltzmann's statistical mechanics and immaculate classical thermodynamics.

The occurrence of the cross-contradiction between the mature math projects of nature creates a shambles of problem situation, the 'Kuhnian crisis.' In social-psychological terms,

> "[Kuhnian] crisis arises at the point when anomaly cannot be localized, when scientists sense that they no longer quite know how to go on becomes more general. Crisis is debilitating precisely because it cannot be localized as the falsification of some beliefs; it is instead the *incoherence* of research practice. The alternative to normal scientific practice being 'in the truth' is not error, but *confusion*" (Rouse 1914, p.5).

For instance, the phlogiston chemistry decisively failed as a practical tool for further exploring and coping with the relentless proliferation of sundry 'airs.' For the working 'normal scientist.' an inability to proceed with research in a coherent way must eventually be an inseparable objection to any theoretical program.

9.The profound contradictions between 'math projections of nature' can be effectively eliminated in a more general (global) theory. The hallmarks of global theory and the reductionist and synthetic ways of its construction, as well as the inexorable stages of its thorny contrivance, are scrutinized. It is contended that the unfolding the cross-contradictions constitute the bifurcation points where the influence of socio-cultural factors is the strongest. This suggests a rather conservative though flexible mode of a scientific upheaval. Seen that way, scientific change is not substantially different from other types of socio-cultural changes.

10. A synthetic global theory is groped in reality by trial and error. Quite contrary to a reductionist global theory (with stiff rules of construction), a synthetic mature theory is contrived gradually. In the course of the construction, intermediate hypotheses are discreetly proposed that happen to possess obvious opportunities of providing constant empirically progressive problem shift in comparison with reductionist programs.

This is because a basic system of theoretical objects and various systems of derivative objects as well possess two indissoluble assets (Stepin 2005). Firstly, a system of theoretical objects, a theoretical model is a sui generis *representation* of an array of objects under consideration. For instance, the solid basis of classical mechanics meticulously depicts mechanical motions with the help of the ideal model. The model immaculately represents the motions of real bodies by the motions of the 'material points' in the 'inertial systems of reference' under the influence of the 'forces.'

On the other hand, a system of theoretical objects makes up a model that seeks to portray the features of actual *experimental device*. This property permitted Vyacheslav S. Stepin to call a basic system 'A Fundamental Theoretical Scheme' (FTS) and a derivative system 'A Partial Theoretic Scheme' (PTS). For instance, the FTS of classical mechanics that represents mechanical motion by the motion of a material point (in an inertial frame of reference under the influence of the forces) features a *gedankenexperiment.* The latter catches the hallmarks of multifarious mechanical experiments with real bodies. The secure basis of Newtonian mechanics is a generalization of practical operations with real bodies moving along the inclined planes, operations with oscillating pendulums, with collisions of billiard balls, and so on. In particular, the PTS of small-oscillations constitutes a model describing the various features of the experiments with oscillations of real pendulums, real stretched strings, etc.

Radical elimination of the cross-contradictions is chiefly possible in such global theory that tentatively contains all the cross-basises as *different* systems of derivative objects. In my humble opinion, only in synthetic global theory can the global theoretical scheme constitute a model that aptly combines all the salient features of idealized experiments with the abstract objects of *both* cross-theories. It is exhibited that the creation of the global synthetic theory is accompanied by the constant empirically progressive problem shift.

11. It is elicited that the insertion of the crossbreeds engenders contradictions in all mature cross-theories and forces them to accommodate each other, to 'interpenetrate' and to 'enrich' each other. Hence the model proposed can render a *mechanism* of simultaneous resolute transformation of 'old' classical paradigms. The hallmark of the mechanism is the construction of the crossbred systems from the basic theoretical objects of 'old' mature theories. These crossbred systems are subsequently generalized to erect a structure of a novel mature theory. A mature physical theory can be empirically successful if it represents a kind of 'ein entwurf' (a draft) that can be explicated by the results of new experiments that increase its empirical content.

12. The model is illustrated in a walk chiefly regarding intricate physics of the early XX-th century, the four 'old' mellow programs in the case being Maxwellian electrodynamics, Newtonian mechanics, Boltzmann's statistical mechanics and classical thermodynamics, and to general relativity genesis. Enticing modern examples, referring to electroweak theory contrivance and general relativity and quantum field theory craving reunion, are also involved.

Yet further effective application of the lucid model for rational reconstructions of new theory-change episodes demands the consideration of the notorious 'incommensurability problem,' coined by T. Kuhn and P. Feyerabend. Namely that is done in the third chapter.

13. In **chapter III** it is contended that one of the promising directions for solving the problem of incommensurability of paradigms is linked with the trend of advancement of neo-Kantian epistemology, embodied by the writings of Ernst Cassirer. According to Cassirer, they are the statements fixing connections and relationships between mathematical ideal constructs that mould a 'neutral language' that can serve as a firm ground for comparing the 'old' and 'new' paradigms.

14. To perceive it, one should take into account that modern science arose only in the XV - XVII centuries. It happened when two diverse research traditions that had previously developed relatively independently of each other encountered and began to 'interpenetrate and actively interact' in Western European culture.

The first tradition (Bacon) is the tradition of experimental natural science, according to which all scientific knowledge arises out of experience. The second tradition (Descartes) is the tradition of mathematically immaculate reasoning, according to which every scientific statement must be proven, i.e. in logically impeccable way derived from the system of self-evident axioms.

Intense interaction of these traditions resulted in the construction of systems of ideal mathematical objects. As a result, in place of single 'values of the series,' established by observation or measurement, one puts 'limiting values towards which the series as a whole move.' Natural science theories provide illustrative examples of the application of this constructive method of 'passing to the limit': a perfectly elastic body, an ideal gas, an incompressible fluid, a perfect circular process, etc. Classical mechanics itself describes the strict laws of motion only through the use of the ultimate idea of a 'material point.'

15. Generalizing numerous examples described in detail in the works of Heinrich Hertz, Cassirer concludes that the fundamental concepts of the natural sciences, and especially the basic concepts of mechanics and electrodynamics, are by no means *copies* of things directly given in the experience. On the contrary, they are the necessary 'constructive blueprints' of physical thought. The theoretical significance of these projects is primarily determined by the extent to which the consequences derived from them coincide with the experimental results obtained.

In accordance with the entrenched Kantian tradition, Cassirer's epistemology proceeds from the fact that modern science categorically rejects natural philosophical claims to penetrate 'inside' nature, implying its last substantial foundations. The development of science appears in the Cassirer version of neo-Kantian epistemology as follows.

(I) First, a careful differentiation is made of the individual spheres of perception offered to us by empirical contemplation. In each of them, 'fluid transitions' are replaced by precise, well-fixed *numerical* definitions. But already at this stage, in order to enter the theoretical schemes of optics, thermodynamics, etc., the sensory content must go through the transformations described above, when certain sensations are replaced in the primary schemes by certain limits of mathematical sequences or, in Russian vocabulary (Stepin) 'particular theoretical schemes.'

(II) But inevitably the second, more significant stage comes. "The genuine and much more difficult task of physics is not to pass from sensible 'qualities' to precisely defined quantities. Rather, after the implementation of this preparatory work, she raises the really fundamental question for her about *the functional connection* and connection of individual qualitative areas. They should not be made out in simple coexistence and succession, but they should be thought of as a law-like and law-governed *unity*" (Cassirer 1923, p.340).

Eventually,

> "the consistent development of a genuinely physical form of thinking and its symbolism gave rise to a [neutral!] language that created the possibility of such an unheard-of expansion, which disclosed that the boundaries of sensation are merely random 'anthropomorphic' ones" (Cassirer 1923, p. 347).

16. Progress in the development of physics does not at all consist in a deeper immersion into the essence of the processes under consideration, expressed in discovering, finally, the true 'substance' from which all things are made. It consists in combining various physical theories, in bringing the integrity of natural phenomena under one universal rule that establishes connections and relationships between various sections of physics.

And the inherent result of this unification should be a new system of relations describing the connections between already known idealizations of physical processes. Thus, the process of unification of scientific theories turns out to be inextricably linked with another deeper process expressing the progress of scientific knowledge. As a result of the unification of physical theories

> "With the individual peculiarity of impressions, its inner heterogeneity has also disappeared, so that regions that, from the point of view of sensation, are *completely incomparable* with each other, can now be understood as interconnected members of the same general plan. Only in this lies the special value of scientific constructive building; in it turns out to be connected by means of continuously stretching intermediate logical terms that which in the first naive view is alien and unrelated to each other" (Cassirer 1923, p.352, my italics).

17. A vigorous case study of the genesis and functioning of a neutral mathematical language related to the Maxwellian synthesis of optics and electromagnetism is provided in **chapter III.** To begin with, one should take Ludwig Boltzmann's comments on Maxwell's works more seriously. In his comments on Maxwell's electrodynamics papers, the founder of statistical mechanics had stressed that many Maxwell's works but especially his early electrical papers "were not properly understood." These works "written according to the *long-term plan*" disclose that their author "was *as mastermind in theory of knowledge as he was in the field of theoretical physics*."

Maxwell's genuine neo-Kantian epistemology sprung out from an intention to find a fruitful compromise between the extremes of Kantian relativism and Scottish 'common sense realism.' It was an indispensable part of his ambitious unification of optics and electromagnetism design.

The lessons properly taught by Kantian epistemology boil down to this.

(i) "The principle of relativity of scientific truth." (Note that even the examples of the analogies are taken by Maxwell from the '*Prolegomena*').

(ii) 'Theory laidenness of observation.'

> "The dimmed outlines of phenomenal things all *merge into one another* unless we put on the focusing glass of theory, and screw it up sometimes to one pitch of definition and sometimes to another, so as to see down into different depths through the great millstone of the world (Maxwell; quoted from Campbell & Garnett 1882, p. 125).

(iii) 'Unity of the laws of Nature.' In nature all the phenomena are tightly *interconnected* and merge into one another. All the differences in theoretical approaches are due to the fact that their authors focus on the multifarious facets and different levels of the phenomena investigated. Hence a theoretician's task is to provide the special notions to cover the various domains of experience.

> "It is hard work *grinding out 'appropriate ideas,'* as Whewell calls them. I think they are coming out at last, and by dint of knocking them against all the facts and half-digested theories afloat, I hope to bring them to shape, after which I hope to understand something more about inductive philosophy that I do at present.
>
> I have a project of *sifting* the theory of light and making everything stand upon definite experiments and definite assumptions, so that things may not be supposed to be assumptions when they are either definitions or experiments" (Maxwell; quoted from Campbell & Garnett, 1882, p.112).

At first the 'appropriate ideas' are vague and dim; yet in the long run they are 'grinded out' by knocking them with the 'facts,' as well as other theories. Though theoretician's task is not only to introduce and furnish the subtle notions 'reflecting' the different facets of the phenomena under consideration, but also to ***unify*** the notions in effective synthesis.

The stages of such a synthesis are highlighted in Maxwell's chef-d-euvre "*Hermann Ludwig Ferdinand Helmholtz.*"

> "Hence the ordinary growth of human knowledge is by accumulation round a number of *distinct* centers. The time, however, must sooner or later arrive when two or more

departments of knowledge can no longer remain independent of each other, but *must* be fused into a consistent whole. But though men of science may be profoundly convinced of the necessity of such a fusion, the operation itself is a most arduous one. For though *the phenomena of nature are all consistent with each other*, we have to deal not only with these, but with the hypotheses which have been invented to systematize them; and it by no means follows that because one set of observers have labored with all sincerity to reduce to order one group of phenomena, the hypotheses which they have formed will be *consistent* with those by which a second set of observers have explained a different set of phenomena.

Each science may appear tolerably consistent within itself, but before they can be combined into one, each must be stripped of the daubing of untempered mortar by which its parts have been prematurely made to cohere.

Hence the operation of fusing two sciences into one generally involves much criticism of established methods, and the explosion of many pieces of fancied knowledge which may have been long held in scientific reputation" *(Maxwell [1877], 1890, p.592).

18. It should be stressed that Maxwell's pivotal method of the '*physical analogies*' also sprung out of Kantian epistemology. The method decidedly rejects the 'ontological' approaches striving for the 'essences' of electrical and magnetic phenomena and proclaiming that 'in reality' electricity and magnetism are sheer 'fields' and not 'action at a distance' phenomena, or vice versa. Maxwell's proposal is to take Faraday's lines of force as a kind of inscrutable tubes filled with ideal incompressible fluid.

"I propose then, […]; and lastly to show how by an extension of these methods, and the introduction of another idea due to Faraday, the laws of the attractions and inductive actions of magnets and currents may be clearly conceived, without making assumptions as to the *physical nature* of electricity, or adding anything to that which has been already proved by experiment.

> By referring everything to the purely geometrical idea of the motion of an imaginary fluid, I hope to attain generality and precision, and to avoid the dangers arising from a premature theory professing to explain the cause of the phenomena" (Maxwell [1856], 1890, p. 159).

It is crucial for a neo-Kantian that this incompressible fluid has nothing to do with experimentally given reality. The single constrain on the concocted models consists in the demand that the mathematical constructs should not contradict each other. In all the other delicate matters the physical analogies method admits an unlimited freedom of imagination. Even the relentless conservation laws can be broken down!

> "There is nothing self-contradictory in the conception of these sources where the fluid is created, and sinks where it is annihilated. The properties of the fluid are at our disposal, we have made it incompressible, and now we suppose it produced from nothing at certain points and reduced to nothing at others" (Maxwell [1856], 1890, p. 162).

And in the other parts of the paper Maxwell exhibits the ways by which the idea of incompressible fluid motion can be applied to the sciences of statical electricity, permanent magnetism, magnetism of induction, and uniform galvanic currents. It should be stressed that the core element of his innovations consisted in the construction of '*neutral language*' for description and comparison of the consequences from the rival theories and competing theoretical models. Maxwell's 'neutral language' was not Carnap's and Reichenbach's 'observation language' directly springing out from the 'protokolsatze' generalizations.

In order to compare and to integrate in a theoretical scheme lacking contradictions all the results of dissent experiments carrying the footprints of multifarious theoretical languages, it is necessary to construct an *artificial* theoretical language equally distant from the languages of theories under comparison. This stupendous language turned out to be the solid state mechanics (with hydrodynamics as its inherent part). Maxwell's ultimate aim was to rewrite all the known empirical and theoretical laws of electricity and magnetism applying the neutral language and then to compare them in order to contrive a system without contradictions.

Again and again he has to emphasize that

> "the author of this method of representation does not attempt to explain the *origin* of the observed forces by the effects due to these strains in the elastic solid, but makes use of the mathematical *analogies* of the two problems to assist the *imagination* in the study of both" (Maxwell [1861], 1890, p.163; my italics).

19. In **chapter IV** the genesis of the so-called 'old quantum theory' represented by Planck's startling 1897-1906 papers is elicited. It is manifested that the primary quantum-theoretical schemes were punctiliously constructed as genuine 'crossbreed' ones composed of ideal models and laws of Maxwellian electrodynamics, Newtonian mechanics, statistical mechanics, and classical thermodynamics. In this pioneering medley, Ludwig Boltzmann's seminal ideas and subtle technique turned out to be crucial. Forsooth, deriving his blackbody radiation law, Max Planck had to take the experimental evidence into account. Yet it spurred him not 'to deduce from phenomena' punctiliously but to involve more theory instead. The blackbody solid experiments incited Planck to apply the statistical technique for proper arranging the radiation data in increasing portions. It is contended that Planck's startling discovery was by no means accidental. Planck's hybrid and intricate theories epitomized the commencement of an ambitious program of Maxwellian electrodynamics and statistical mechanics effective reconciliation.

20. It is contended that Einstein's 'annus mirabilis' constituted an acme of the second scientific revolution. To fathom in what felicitous ways Einstein's 1905 writings hang together one is bound to pay special tribute to his longed strive for unity evinced in incessant attempts to coordinate the profound research traditions of classical physics. Nevertheless, Einstein's craving efforts sprung out of Max Planck's pioneering attempts to comprehend electromagnetic phenomena through the lenses of conceptual structures of thermodynamics and statistical mechanics. It was Planck, who clearly realized the cross-contradiction between 'the physics of material bodies and the physics of the ether' and outlined the first sketch of its withdrawal: the paradigms 'must be modified to remain compatible.' And it was Planck who took the first step in modifying the physics of the ether and contending that 'not only matter itself but also the effects radiated from matter' possess discontinuous properties. They can be characterized by a new natural constant: the elementary quantum of action. Einstein's indispensable part consisted in that he took the next step in modifying the second component of the encounter – the physics of material bodies. Forsooth, "the modification in mechanics due to the principle of relativity is the introduction of a new universal constant, entirely foreign to classical mechanics, namely the velocity of light in an absolute vacuum" (Planck). Einstein's startling light quanta hypothesis and distinctive special theory of relativity turn out to be mere milestones of the unwinding of Maxwellian electrodynamics and statistical thermodynamics reconcilement research program. The notorious conception of luminiferous ether was a substantial obstacle for Einstein's wayward statistical thermodynamics in which the pivotal lead was played by flagrant light quanta paper. Herewith Einstein was aware that his enticing light quanta hypothesis was too audacious to be taken literally. Hence he laid out his own version of 'electrodynamics of moving bodies' in a markedly Machian/Duhemian, phenomenological way, and the basic relativity postulate was disposed of as originating from unfathomable and subconscious 'instinctive knowledge.' Forsooth, in his relentless criticism of the moldy research traditions of classical physics David Hume and Ernst Mach. influenced Einstein. However, when related to *creative* momenta, his 1905 reconcilement *modus operandi* was initiated by Mach's pre-eminent 'principle of economy of thought.' Yet the principle was taken in the subtle context of Stevin-Mach 'instinctive knowledge' wayward doctrine and with promising inspirations of the regulative spirit of Kantian and Duhemian non-representationalist epistemologies. It is manifested that for Einstein Duhem's confounded conventionalism was an indispensable bridge connecting and accommodating two quite disparate subjects of

Machian crude empiricism and Kantian academic apriorism and transcendentalism. It constituted an indispensable means of saving special relativity postulates from scarce criticism for their inseverable connection with the foolhardy light quanta ideas.

21. The aim of the **chapter IV** is to take a step towards a complete description of Special Theory of Relativity genesis and acceptance, bringing some additional light on the intertheoretic relations between STR and other physical theories of the day. It should be stressed that the approach proposed enables to explain *why* classical mechanics and classical electrodynamics were "refuted" almost simultaneously or, in more suitable terms, why did the quantum and the relativistic revolutions both took place at the beginning of the XX-th century. It is exhibited that the Quantum and the Relativistic revolutions were simultaneous since they had the common origin - the profound clash between the fundamental theories of the second half of the XIX-th century that constituted the 'body' of Classical Physics. The revolution's most dramatic turning point was Einstein's 1905 light quantum paper, that laid the foundations of the Old Quantum Theory and influenced the fate of the STR too.

22. What was Dirac's own philosophy of science that allowed him to unify such diverse realms as 'relativity' and 'quanta'? Regrettably, Dirac's implicit philosophy of science has to be piecemeally reconstructed from his stout papers and scant writings on general questions. The aim of **the chapter V** is to exhibit that, despite the relentless Dirac's remarks on the notorious role of philosophy, one can trace a *latent* influence of Husserlian epistemology through Arthur Eddington's and Hermann Weyl's whimsical mathematical writings embracing their paradigmatic efforts to merge gravitation with electromagnetism.

23. Even though the existence and functioning of a conscious metaphysical 'hard core' is extremely doubtful in the Dirac peculiar case, one should admit the existence of the 'positive heuristic' of Dirac's SRP, which, without any significant changes, permeated *all* his scientific creativity. Dirac's positive heuristic of the following principles.

(I) As the *first principle*, one can single out Dirac's beliefs about the smooth, evolutionary development of science and the need for prudent and gradual (piecemeal) changes in the content of mature scientific theories.

(II) *The Principle of Unity of Science* asserted that "all of Nature is amenable to the same kind of theoretical treatment."

(III) The *third* in a row (but not in importance!) - put forward by Dirac as a *pivot* - the so-called '*principle of mathematical beauty*.'

(IV) The *fourth* principle of the Dirac heuristic is closely related to those considered above: they are Einstein's theories of relativity (STR and especially GTR) that constitute a sublime *ideal*, and unsurpassed specimen of mathematical beauty.

(V) Interpretation of the new mathematical features in terms of physical entities is provided by "*Eddington's Principle of Identification*."

Eliciting the principle, Eddington refers primarily to the mathematical identification of the vector and tensor structures of his abstruse 'infinitesimal world geometry' with the subtle structures of gravity and electromagnetism (i.e. with the structures of unified theories) pioneered by Hermann Weyl (1919). Weyl's "world geometrical" framework for the new physics turned out to be the result of a second-order 'eidetic analysis,' carried out by phenomenological reflection upon the differential geometrical framework of the GTR. Eddington, in turn, appeared to be heretical enough to accept Weyl's generalization of Einstein's GTR and to generalize it further, for philosophical reasons essentially similar to Weyl's.

(VI) The next operative heuristical principle is epitomized by Weyl's unified theory of electricity and magnetism published in 1918. Dirac was indirectly influenced by the Weyl-Husserl epistemology not only via Eddington's numerous writings but through his first-hand acquaintance with Weyl's unified theory of electricity and gravitation.

(VII) The most controversial component of Dirac's heuristic is his '*profound intuition*' commonly contrasted to 'logical clarity in the presentation.' It was always Dirac's ideal, in formulating a physical theory, to be able to construct or to discover a 'reasonable mathematical scheme' and to interpret the equations at hand most directly and naturally. The scheme must be mathematically well-defined and self-consistent, and in agreement with well-established principles; yet these criteria are quite insufficient even for choosing a definite scheme among many. All appearance, to comprehend the problem situation, one should introduce something like Husserl's 'principle of all principles,' extending, in addition to sensory givenness, experience, in the form of the 'given,' to intuition, i. e. to nonsensuous, to 'experience in fantasy' or in imagination.

24. The main stages of the genesis and becoming of any successful mature theory that played a decisive role in the creation of Dirac's masterpiece as are as follows.

(i) The inevitable 'encounter' of stiff and diverse pivotal 'mathematical projections of nature' (STR vs. NQM) and subsequent step-by-step contrivance of a medley of 'hybrid' theoretical schemes. A hybrid theoretical scheme describes its subject area so that there is a pressing need for their synthesis and for constructing a global theoretical scheme with the relations between the theoretical objects related by the global theory basic equations. In the case under consideration – the Dirac synthesis - the hybrid schemes were incarnated by Klein-Gordon-Fock (1926) and Pauli-Heisenberg-Jordan (1926) half-phenomenological theories.

(ii) It commonly turns out that the blunt resolution of the problem situation via elevating the basic objects of any one of the constructed hybrid theoretical schemes on the pedestal of the global theory simply 'does not work.' (Commonly due to explanatory problems of the global reductionist theory in embracing significantly heterogeneous domains of validity). Eventually, the understanding comes that the desired synthesis is possible only via the 'radical transformation' (Dirac), i.e. due to the involvement of some substantially novel branch of mathematics. Its conceptual system should be able to generate a 'neutral language' vocabulary for describing, comparing, and 'reconciling' all the hybrid schemes involved (Cassirer 1910).

In the GTR genesis, the subtle notions of Riemannian geometry and tensor calculus also commenced to function as an efficacious neutral language; withal tensors constitute the geometric objects, the relations between which are preserved in the transition from one non-inertial frame of reference to another.

In the Dirac theory genesis, the role of such a mellow language was played by the odd language of spinor calculus with its Clifford algebra, primarily referred to in the works of W. Pauli in 1926.

(iii) The neutral language obtained is used to generalize all the available hybrid schemes and arrive at the simplest equation embracing all the hybrid regularities encountered (Einstein &Grossman's '*Entwurf*'; Dirac's relativistic electron equation). Such a generalization commonly takes place in one of the encountered hybrid theoretical schemes, and has a *local* character at first, but eventually acquires a *global* significance. The object lesson is the introduction of the four-dimensional function $\psi = \psi\ (x_0, x_1, x_2, x_3)$ by Dirac.

(iv) The decisive stage is taking place. Among the mathematical objects, the relations between which are described by the proposed equations of the global theory, one should discover an object that possesses the properties of the basic object of the global theoretical scheme. Viz., it is necessary to reveal the 'dual' (Eddington) or 'hybrid' (Nugayev) nature of the theoretical object. Viz., to demonstrate that it simultaneously possesses certain 'geometric' properties, transforming itself as a tensor $g_{\mu\nu}$ and, on the other hand, has 'physical' properties also, since it is transformed as an irreducible representation of the Poincaré group, the Lorentz group, etc. This elucidation of the so-called 'sheer physical,' operational sense of mathematical terms should allow comparing the obtained quantitative relationships with experimental data.

Accordingly, in Dirac's relativistic electron theory, the wave functions are described by bispinors ψ, the components of which are complex numbers. Moreover, both in Dirac's theory, as well as in QFT in general, bispinors are connected with such fundamental particles as electrons and even quarks. Under transformations of the Lorentz group, which describe the symmetries of the Minkowski space, the bispinors are transformed as specific irreducible representations of this group - as the set of operators in the Hilbert space. It is the symmetries of flat space-time that famously allow, by Noether's theorem, electrons, and quarks to possess such conserved properties as mass, spin, and charge.

Consequently, ψ - the pivotal mathematical construct of the Dirac equation - has an explicit *dual structure*. On the one hand, it is a typical bispinor. As such, when the coordinate system changes (including rotations in 3-dimensional space and transformations between frames of reference moving rapidly relative to each other), it is transformed not as a vector of ordinary space, but as a mathematical object in a special space called 'internal,' not intersecting with the usual, 'external' space.

But, on the other hand, the same ψ represents, at least in the first approximation, the amplitude of the probability of one electron, associated with the density of the charge and current of the Dirac particle. By the conservation of charge, the quantity that is taken to be the total probability of finding a particle is also conserved. Accordingly, Dirac's theory not only featured the geometric properties of bispinors, but was also able to predict the magnetic moment of an electron, as well as the fine structure of lines in the spectrum of atoms, and, famously, the existence of antiparticles (or positrons).

25. In sum, the successful synthesis consisted in contriving a miraculous equation, the space-time properties of which matched the STR, while the quantum properties reconciled with the NQM transformation properties. Surely, the main asset of Dirac's theory consisted in the theoretical reproduction of the electron spin, which turned out now to be a *strictly relativistic consequence* of his theory.

Just like Einstein generalized the available theoretical schemes using the vocabulary of tensor calculus, Dirac generalized the theoretical schemes of Klein-Gordon-Fock and Pauli-Heisenberg-Jordan using no less powerful and advanced language of the theory of bispinors and Clifford algebra.

26. Finally, the important hallmark of synthesis consisted in the fact that Dirac ascertained the specifics of the GTR synthesis through the peculiar prism of his Cambridge colleague and mentor - Sir Arthur

Eddington, the author of the best (according to Einstein himself) exposition of STR and GTR - monograph '*Mathematical Theory of Relativity.*' Note that for Eddington the mathematical constructs of geometry – all those tensors, vectors, and 'square roots of vectors' - were the *primordial bricks* for stout "World Building."

The world-building is grounded on 'synthesis of representations' (in line with Husserlian notable ideas) in the spirit of the quirk theories of Cartan and Weyl. Therefore, having conceived the methods of constructing dual (hybrid) objects, containing, among other things, recipes for 'interpreting new mathematical properties in terms of physical quantities (using a process similar to Eddington's Identification Principle),' Dirac was in no hurry to reveal the underlying philosophical principles of Eddington's philosophy. The odd principles evoked an extremely negative reaction from the majority of Copenhagen-minded physicists.

REFERENCES

Abiko S. (1991) On the chemico-thermal origins of Special Relativity. *Historical Studies in the Physical Sciences*, vol.22, pp. 1-24.

Abiko S. (2005) The origin and concept of Special Relativity. – In: Mamone Capria (ed.) *Physics before and after Einstein: a historical perspective*. IOS Press, Amsterdam, pp. 71-91.

Abraham, Max, 1912a. Zur Theorie der Gravitation, *Physikalische Zeitschrift*, 13, 1-4.

Abraham, Max, 1912b. Das Elementargesetz der Gravitation, *Physikalische Zeitschrift*, 13, 4-5.

Abraham, Max, 1912c.Der Freie Fall. *Physikalische Zeitschrift*, 13, 310-311.

Abraham, Max, 1912d.Nochmals Relativität und Gravitation. Bemerkung zu A. Einsteins Erwiderung. *Annalen der Physik*, 38:1056-58.

Abraham, Max, 1915. Neuere Gravitationstheories, *Jahrbuch der Radioaktivität und Elektronik*, pp. 470-520.

Agassi, Joseph (1983) The Structure of the Quantum Revolution. *Philosophy of Social Science*, vol. 13, pp.367-81.

Arabatzis Theodore (2017) The Structure of Scientific Revolutions and History and Philosophy of Science in Historical Perspective.. – In: Blum A., K. Gavroglu, C. Joas, J. Renn (eds.) (2017) *Shifting Paradigms. Thomas S. Kuhn and the History of Science.* Proceedings of the Conference "50 years Since Structure: Towards a History of the History of Science," held in Berlin in October 2012.Editor Open Access. Max Planck Institute for the History of Science, pp.191-202.

Ariew, Roger (1989) The phases of Venus before 1610. *Studies in History and Philosophy of Physical Science*, 18:81-92.

Aristotle (2007a) *Physics* (translated by R.P. Hardy and R.K. Goye). eBooks @ Adelaide.

Aristotle (2007b) *Metaphysics* (translated by W.D. Ross). eBooks @ Adelaide.

Audi Robert (ed.) *The Cambridge Dictionary of Philosophy*. Cambridge, Cambridge University Press, 1999. – 1000 p.

Baracca A. (2005) Mechanistic science, thermodynamics and industry at the end of the nineteenth century. In: Mamone Carpia (ed.) *Physics before and after Einstein: a historical perspective.* IOS Press, Amsterdam.

Barker, Peter and Bernard R. Goldstein (2001) Theological Foundations of Kepler's Astronomy. *Osiris, 16,* pp.88-113.

Barker, P. and Vesel M. (2012) Goddu's Copernicus. *Aestimatio*, 9:304-336.

Barnes S. (1977) *Interests and the Growth of Knowledge*. London: Routledge & Kegan Paul.

Bardeen J.M., B. Carter, and S.W. Hawking (1973) "The Four Laws of Black Hole Mechanics." *Communications in Mathematical Physics* 31:161-170.

Barnes, Barry (1982) *T.S. Kuhn and Social Science.* London: Macmillan.

Bechtel, William (1993) Integrating Sciences by Creating New Disciplines. The Case of Cell Biology. *Biology and Philosophy* 8, 277-299.

Beckenstein, Jacob D. (1973) Black Holes and Entropy. *The Physical Review* D7, number 8, pp.2333-2346.

Beller M (2000) Kant's Impact on Einstein's Thought. In: Don Howard and John Stachel (ed.) *Einstein: The Formative Years, 1879-1909*, 83-106, Boston: Birkhauser.

Berger Peter, and Thomas Luckmann.1996. *The Social Construction of Reality: A Treatise in the Sociology of Knowledge.* Garden City, NY: Anchor, 1966.

Biemel Walter (1987) *Martin Heidegger mit Selbstzeugnissen und Bolddokumenten*. Rowohlt, . – 284 Seiten.

Bird, Alexander (2018) Thomas Kuhn. *The Stanford Encyclopedia of Philosophy* (Winter 2018 Edition), Edward N. Zalta (ed.) URL = https://plato.stanford.edu/archives/win 2018/ entries/ thomas - kuhn /

Birrell, N.D., and P.C.W. Davies (1982) *Quantum Fields in Curved Space*. Cambridge: Cambridge University Press.

Bjorken David and Sydney Drell (1998) *Relativistic Quantum Mechanics*. NY: McGraw-Hill Education.

Bloor D. (1991 [1976]) *Knowledge and Social Imagery*, 2nd ed. Chicago: University of Chicago Press.

Bloor, David (1983) *Wittgenstein and the Social Theory of Knowledge*. London: Macmillan

Bloor, David (2017) The Pendulum as a Social institution: T.S. Kuhn and the Sociology of Science. – In: Blum A., K. Gavroglu, C. Joas, J. Renn (eds.) (2017) *Shifting Paradigms. Thomas S. Kuhn and the History of Science*. Proceedings of the Conference "50 years Since Structure: Towards a History of the History of Science," held in Berlin in October 2012.Editor Open Access. Max Planck Institute for the History of Science, pp.235-252.

Blum A., K. Gavroglu, C. Joas, J. Renn (eds.) (2017*) Shifting Paradigms. Thomas S. Kuhn and the History of Science.* Proceedings of the Conference "50 years Since Structure: Towards a History of the History of Science," held in Berlin in October 2012.Editor Open Access. Max Planck Institute for the History of Science.

Bokulich Alisa (2004) Open or closed? Dirac, Heisenberg and the relation between classical and quantum mechanics, *Studies in History and Philosophy of Science,* 35 (3), 377-396.

Boltzmann Ludwig (1895) *Ueber Faraday's Kraftlinien, von James Clerk Maxwell. Herausgegeben von. L. Boltzmann*. Leipzig: W. Engelmann. – 130p.

Boltzmann L. (1905) The Relations of Applied Mathematics. – In: Rogers H.J. (ed.) *Congress of Arts and Science, St. Louis 1904*, vol.1, Boston and New York, Houghton, Mifflin and C., pp.591-603.

Bordoni, Stefano (2013) Widening the boundaries of classical physics: from Einstein's 1909 paper back to late nineteenth-century theoretical physics. In: Isabella Tassani (a curadi*). Oltre la fisica normale.*

Interpretazioni alternative e teorie non standard nella fisica modern, University of Urbino, pp. 25-52.

Born, Max (1971) *The Born-Einstein Letters: Correspondence between Albert Einstein and Max and Hedwig Born from 1916 to 1955.* I. Born translation. N.Y.: Walker and Company.

Brown L.M., Rechenberg H. (1987) Paul Dirac and Werner Heisenberg – a Partnership in Science. – In: Kursunoglu B.N. and Wigner E. P.(eds.) *Paul Adrien Maurice Dirac. Reminiscences about a Great Physicist.* Cambridge University Press, 117-162.

Brush, Stephen (2015) *Making 20th Century Science: How Theories Became Knowledge.* Oxford University Press .

Buchdahl, Gerd (1969) *Metaphysics and the Philosophy of Science.The Classical Origins: Descartes to Kant.* Cambridge, Massachusetts: MIT Press.

Buchwald Jed Z. (2017) Thomas Kuhn. – In: Blum A., K. Gavroglu, C. Joas, J. Renn (eds.) (2017) *Shifting Paradigms. Thomas S. Kuhn and the History of Science*. Proceedings of the Conference "50 years Since Structure: Towards a History of the History of Science," held in Berlin in October 2012.Editor Open Access. Max Planck Institute for the History of Science, pp.151-162.

Burchill G. et al. (1991) *The Foucault Effect: Studies in Governmentality.* Brighton: Harvester.

Burtt Edwin Arthur (2003) *The Metaphysical Foundations of Modern Science*. N.Y.:Dover Publications.

Campbell Lewis, Garnett William (1882). *The Life of James Clerk Maxwell.* L., Macmillan. - 342p.

Cassirer Ernst (1910) *Substanzbegriff und Funktionbegriff: Untersuchungen uber der Grundfragen der Erkenntskritik.* Berlin.

Cassirer, Ernst ([1921], 1923) *Substance and Function and Einstein's Theory of Relativity*, (W.C. Swabey and M.C. Swabey, trans.) Chicago: The Open Court Publishing.

Cassirer Ernst (1923) *Philosophie der symbolischen Formen.* Bd. I. Die Sprache.Berlin.

Cassirer Ernst (1950) *The Problem of Knowledge – Philosophy, Science, and History since Hegel.* New Haven: Yale University Press.

Chalmers A.F. (1975) Maxwell and the displacement current // *Physics Education*, January, pp. 46-49.

Chang Hasok (1995) The Quantum Counter-Revolution: Internal Conflicts in Scientific Change. *Studies in History and Philosophy of Modern Physics*, vol.26, no 2, pp.121-136.

Cheng Hasok (2022) *Realism for Realistic People. A New Pragmatist Philosophy of Science.* Cambridge University Press.

Chaunu, Pierre (1984) *La Civilisation de l'Europe Classique*. Paris: Les Editions Arthaud.

Copernicus, Nicholas ([1543], 1972) *On the Revolutions of the Heavenly Bodies* (translated by Edward Rosen). London: Macmillan.

Collins, Harry (2017) The Notion of Incommensurability. – In: Blum A., K. Gavroglu, C. Joas, J. Renn (eds.) *Shifting Paradigms. Thomas S. Kuhn and the History of Science*. Proceedings of the Conference "50 years Since Structure: Towards a History of the History of Science," held in Berlin in October 2012.Editor Open Access. Max Planck Institute for the History of Science, pp.253-258.

Darrigol O. (2001) *Electrodynamics from Ampere to Einstein.* Oxford: Oxford University Press.

De Broglie Louis (1962) Mysterious constant h - Max Planck's celebrated discovery. - In: *Following the Paths of Science*. Moscow, Izdatelstvo Inostrannoy Literaturi (Foreign Literature Publishing House), pp.139-146 (in Russian).

Derrida, J. (1970) Structure, sign and play in the discourse of the human sciences. In *The structuralist controversy: The language of criticism and the sciences of man,* edited by R. Macksey and E. Donta. Baltimore: Johns Hopkins University Press.

De Witt, B.S. (1975) "Quantum-Field Theory in Curved Spacetimes." *Physics Reports* C.19: 297-357.

De Witt, B.S. (1980): "Quantum Gravity: the New Synthesis." In *Quantum Gravity 2, 2nd Oxford Symposium on Quantum Gravity*, edited by C.J. Isham, Roger Penrose and Dennis William

Dirac Paul (1927) "The Physical Interpretation of the Quantum Dynamics," *Proceedings of the Royal Society (London)*, 113, 621-641.

Dirac Paul (1928) The Quantum Theory of the Electron. *Proceedings of the Royal Society (London)*, 610-624.

Dirac Paul (1931) Quantized Singularities in the Electromagnetic Field. *Proceedings of Royal Society, London*, A133, 60-72.

Dirac Paul (1938-1939) The Relation Between Mathematics and Physics. *Proceedings of the Royal Society, Edinburgh*, A 59, 122-129.

Dirac P. A. M. (1963) "The evolution of the physicist's picture of nature." *Scientific American*, 208 (May), 45-53.

Dirac P. A. M. (1966) *Lectures on Quantum Field Theory*. NY: Belfer Graduate School of Science/ Academic Press.

Dirac P. A. M. (1969) Can equations of motion be used? – In: T. Gudehus, G. Kaiser, and A. Perlmutter (eds.) *Fundamental Interactions at High Energy*. NY: Gordon and Breach, 1-13.

Dirac P.A.M. (1973a) Long-range forces and broken symmetries. *Proceedings of the Royal Society*, London, A-333, January 15, 403-418.

Dirac P.A.M. (1973 b) Fundamental constants and their development in time. – In J. Mehra (ed.) *The Physicist's Conception of Nature*. Dordrecht: Reidel.

Dirac P.A.M. (1979) Der Zeittest. *Naturwissenschaftliche Rundschau*, 33: 353-356.

Dirac P.A.M. (1980) Why we believe in the Einstein theory. – In: B. Gruber and R.S. Millman (eds.) *Symmetries in Science*. NY: Plenum Press, pp. 1-11.

P.A.M. Dirac ([1930], 1981) *The Principles of Quantum Mechanics*. International Series of Monographs on Physics. (Oxford: Clarendon Press, 4th edition).

Dmitriev I.S. (2019) I didn't really think about it. *Discourse*, vol.5, no. 5, pp.5-19.

Drake, Stillman (1957) *Discoveries and Opinions of Galileo*. N.Y.: Dover Publications.

Dugas Rene (1959): *La theorie physique au sens de Boltzmann et ses prolongements modernes*. Paris, Editions de Griffon.

Duhem, Pierre ([1905], 1954) *The Aim and Structure of Physical Theory*. Princeton, New Jersey: Princeton University Press.

Eddington, Arthur (1920) *Space, Time and Gravitation. An Outline of General Relativity Theory*. Cambridge: Cambridge University Press.

Eddington, Arthur (1921) *The Mathematical Theory of Relativity*. The University Press.1921

Eddington A.S. ([1928], 1948) *The Nature of the Physical World*. Cambridge University Press, 1948.First edition: November 1928.

Edgerton S. (1980) The Renaissance artist as a quantifier. In *The Perception of Pictures*, edited by M.A. Hagen, vol.1. New York: Academic Press.

Einstein A (1902) Über die thermodynamische Theorie der Potentialdifferenz zwischen Metallen und vollständig dissozierten Lösungen ihrer Salze, und eine elektrische Methode zur Erforschung der Molekularkräfte. *Ann Phys* 8: 798-814. English translation by Anna Beck in: *The Collected papers of Albert Einstein*. Vol. 2. The Swiss years: writings, 1900 – 1909. Princeton, New Jersey: Princeton University Press, 1989, 12-29.

Einstein A (1903) Eine Theorie der Grundlagen der Thermodynamik. *Ann. Phys.* 11: 170-187. English translation by Anna Beck in: *The Collected Papers of Albert Einstein*. Vol. 2. The Swiss years: writings, 1900 – 1909. Princeton, New Jersey: Princeton University Press, 1989, 48-67.

Einstein A (1904) Zur allgemeinen molekularen Theorie der Warme. *Ann. Phys.* 14: 354-362. English translation by Anna Beck in: *The Collected Papers of Albert Einstein*. Vol. 2. The Swiss years: writings, 1900 – 1909. Princeton, New Jersey: Princeton University Press, 1989, 68-77.

Einstein A (1905a) Über einen die Erzeugung und Verwandlung des Lichtes betreffenden heuristischen Gesichtspunkt. *Ann Phys* 17 :132-48. English translation by Anna Beck in: *The Collected Papers of Albert Einstein*. Vol. 2. The Swiss years: writings, 1900 – 1909. Princeton, New Jersey: Princeton University Press, 1989, 86-103.

Einstein A (1905b) Eine neue Bestimmung der Moleküldimensionen. University of Zurich Dissertation. English translation by Anna Beck in: *The Collected Papers of Albert Einstein.* Vol. 2. The Swiss years: writings, 1900 – 1909. Princeton, New Jersey: Princeton University Press, 1989, 104-122.

Einstein A (1905c) Über die von der molekularkinetischen Theorie der Wärme geforderte Bewegung von in ruhenden Flüssigkeiten suspendierten Teilchen. *Ann Phys.* vol. 17. English translation by Anna Beck in: *The Collected Papers of Albert Einstein.* Vol. 2. The Swiss years: writings, 1900 – 1909. Princeton, New Jersey: Princeton University Press, 1989, 123-134. Einstein A. (1905d) Zur Elektrodynamik bewegter Körper. *Ann Phys* 17: 891-921.English translation in: *The Principle of Relativity*, Dover, New York, 1923. See also the translation of Anna Beck in: *The Collected Papers of Albert Einstein.* Vol. 2. The Swiss years: writings, 1900 – 1909. Princeton, New Jersey: Princeton University Press, 1989, 140 - 171.

Einstein A. (1905e) Ist die Trägheit eines Körpers von seinem Energienhalt abhängig? *Ann Phys* 18:639-641. English translation of Anna Beck in: *The Collected Papers of Albert Einstein.* Vol. 2. The Swiss years: writings, 1900 – 1909. Princeton, New Jersey: Princeton University Press, 1989, 172 - 174.

Einstein A. (1907) Über das Relativitätsprinzip und die aus demselben gezogenen Folgerungen. *Jahrbuch der Radioaktivität und Elektronik*: 4, 411-462. English translation of Anna Beck in: *The Collected Papers of Albert Einstein.* Vol. 2. The Swiss years: writings, 1900 – 1909. Princeton, New Jersey: Princeton University Press, 1989, 252-311.

Einstein A. (1909) Über die Entwicklung unserer Anschauungen über das Wesen und die Konstitution der Strahlung. *Physikalische Zeitschrift* 10: 817-25. English translation of Anna Beck in: *The Collected Papers of Albert Einstein.* Vol. 2. The Swiss years: writings, 1900 – 1909. Princeton, New Jersey: Princeton University Press, 1989, 379 - 394.

Einstein A. (1910) Principé de relativité et ses consequencé dans la physique modern. *Archives des Sciences Physique et Naturelles* 29:125-44.

Einstein, Albert (1912 a) Zur Theorie des statischen Gravitationfeldes. *Annalen der Physik*, 38, 443-458.

Einstein, Albert (1912b) Relativität und Gravitation. Erwiderung auf eine Bemerkung von M. Abraham, *Annalen der Physik*, 38, 1059-1064.

Einstein, Albert and Grossmann, Marcel (1913a) Entwurf einer verallgemeinerten Relativitätstheorie und Theorie der Gravitation, *Zeitschrift fur Mathematik und Physik*, 62, 225-261.

Einstein, Albert (1913b) Physikalische Grundlagen einer Gravitationtheorie, *Naturforchende Gesellchaft*, Zurich, Vierteljahrsschrift, 58, 284-290.

Einstein, Albert (1913c) Zum gegenwartigen Stande des Gravitationsproblems, *Physikalische.Zeitschrift*, 14, 1249-1262. Reprinted in: Jürgen Renn, ed., *The Genesis of General Relativity*. Vol.3.Gravitation in the Twilight of Classical Physics: Between Mechanics, Field Theory and Astronomy. (Dordrecht: Springer, 2007).

Einstein, Albert and Fokker, Adrian (1914) Die Nordströmsche Gravitationtheorie vom Standpunkt des absoluten Differentialkalkulis, *Annalen der Physik*, 44, 321-28.

Einstein, Albert (1915) "Die Feldgleichungen der Gravitation," *Preussiche Akademie der Wissenschaften, Sitzungberichte* 48, 844-847.

Einstein A. (1918) Motive des Forschens, In: *Zu Max Planks Sechszigstem Geburtstag*. Ansprachen, gehalten am 26 April 1918 in der Deutschen Physikalischen Gesellchaft, 29-32, Karlsruhe: Muller,

Einstein A. (1949a) Autobiographical Notes. In *Albert Einstein: Philosopher-Scientist*, ed. P.A. Schilpp, vols 1-2, 1-14, Evanston, IL,

Einstein A. (1949b) Remarks Concerning the Essays Brought together in this Co-operative Volume. In: *Albert Einstein: Philosopher-Scientist*, ed. P.A. Schlipp, vols 1-2, 665-688, Evanston, IL.

Einstein A. (1954) Relativity and the Problem of Space.In: *Ideas and Opinions*, 360-370, New York: Crown Publishing Inc.

Einstein A. ([1933], 1954) On the Method of Theoretical Physics. Reprinted in: *Ideas and Opinions*, 277-283. New York: Crown Publishing Inc.

Einstein A. (1944 / 1954) Remarks on Bertrand Russell's Theory of Knowledge. In: *The Philosophy of Bertrand Russell.* ed. Paul Arthur Schilpp, Evanston, Illinois: The Library of Living Philosophers, 278-292. Reprinted in Ideas and Opinions, 18-24, New York: Crown Publishers Incorp.

Einstein A. ([1940], 1954) The Fundaments of Theoretical Physics, In: *Ideas and Opinions*, 323-330, New York: Crown Publishing Inc. .

Einstein, Albert ([1936], 1960) Physics and Reality. *The Journal of the Franklin Institute* 221, no 3 (March). Reprinted in: Albert Einstein. *Ideas and Opinions*. Translated by Sonja Bargmann. N.Y.: Crown Publishers, 1960, 290-323.

Albert Einstein / Michele Besso (1972) *Correspondence 1903-1955*. Paris: Hermann.

Einstein A. (1987) *The Collected Papers of Albert Einstein.* Vol. 1. The Early Years, 1879-1902. ed. John Stachel. Princeton: Princeton University Press.

Einstein A. (1989) *The Collected papers of Albert Einstein.* Vol. 2. The Swiss years: writings, 1900 – 1909. Princeton, New Jersey: Princeton University Press.

Einstein, Albert (1992) *The Collected Papers of Albert Einstein.* Vol.4. The Swiss Years: Writings, 1912-1914. Martin J. Klein, A. J. Kox, Jurgen Renn and Robert Schulmann, eds., Princeton: Princeton University Press .

Albert Einstein / Mileva Marić (1992) *The Love Letters*. Ed. and with an introduction by Jürgen Renn and Robert Schulmann. Transl. by Syawn Smith. Princeton, NJ: Princeton University Press.

Einstein, Albert (1993) *The Collected Papers of Albert Einstein.* Vol.5. The Swiss Years: Correspondence, 1902-1914. Martin J. Klein, A. J. Kox and Robert Schulmann, eds., Princeton: Princeton University Press.

Elkana, Yehuda (1994) *Essays on the Cognitive and Political Organization of Science*. Max Planck-Institut fur Wissenschaftsgeschichte.

Engler Fynn Ole and Jurgen Renn (2017) Two Encounters. . – In: Blum A., K. Gavroglu, C. Joas, J. Renn (eds.) (2017) *Shifting Paradigms.*

Thomas S. Kuhn and the History of Science. Proceedings of the Conference "50 years Since Structure: Towards a History of the History of Science," held in Berlin in October 2012.Editor Open Access. Max Planck Institute for the History of Science, pp.139-148.

Everett, Jonathan, 2015. The constitutive a priori and the distinction between mathematical and physical possibility. *Studies in History and Philosophy of Modern Physics*, 2015, 52: 139-152.

Feinberg Andrew, Leiss William (2007) Introduction to: *The Essential Marcuse: Selected Writings of Philosopher and Social Critic Herbert Marcuse*. Beacon Press, pp. vii-xliii.

Feyerabend, Paul. ([1975], 2010) *Against Method*.4th ed. New York, NY: Verso Books.

Fitzpatrick, Richard (2010) *A Modern Almagest. An Updated Version of Ptolemy's Model of the Solar System.* Lulu Enterprises.

Fölsing, Albrecht (1997) *Albert Einstein: A Biography*. Translated by Ewald Osers. New York: Viking.

Foucault, Michel ([1975], 1977) *Discipline and Punish: The Birth of the Prison*. NY: Random House.

Foucault, Michel (1980) *Power/Knowledge. Selected Interviews and Other Writings*, 1972-1977. Edited by Colin Gordon. NY: Pantheon Books.

Foucault M. (1988) *Aesthetics, Method, and Epistemology*. The New Press, New York.

Foucault, Michel ([1963], 1994) *The Birth of the Clinic*. An Archeology of Medical Perception. Translated from the French by A.M. Sheridan Smith. NY: Vintage Books.

Foucault, Michel ([1966], 1994) *The Order of Things: An Archeology of the Human Sciences*. NY: Vintage Books.

Foucault, Michel ([1969], 2002) *The Archaeology of Knowledge*. Translated from the French by A.M. Sheridan Smith. London and New York: Routledge.

Foucault, Michel ([1961], 2009) *History of Madness*. Translated by J. Khalfa. NY: Routledge. Foucault, Michel. [1976-2021]. *The History of Sexuality*. Four volumes. (1979) [1976] Vol.1. *An Introduction.*

London: Allen Lane. (1992) [1984] Vol.2.*The Use of Pleasure*. London: Penguin Books. (1990) [1984] Vol.3.*The Core of the Self.* London: Penguin Books. (2021) [2021] Vol.4.*Confession of the Flesh*. London: Penguin Books.

Frede, Dorothea (2006) Heidegger's Project: the question of being. – In: *The Cambridge Companion to Heidegger* (ed. by Charles B. Guignon). Cambridge University Press, pp. 42-81.

Fuller, Steve (2003) *Kuhn vs. Popper: The Struggle for the Soul of Science*. Cambridge: Icon Books.

Fuller, Steve (2020) Permanent Revolution in Science: a Quantum Epistemology. *Philosophy of the Social Sciences*, pp. 1-10.

Fulling, S.A. (1973) Nonuniqueness of Canonical Field Quantization in Riemannian Space-Time. *The Physical Review* 7: 2850-2861.

Galileo, Galilei ([1623], 1957) *The Assayer* (translated by Stillman Drake). New York: Doubleday.

Galileo, Galilei ([1632], 2001) *Dialogue Concerning the Two Chief Systems – Ptolemaic and Copernican* (translated by Stillman Drake). New York: The Modern Library.

Galison, Peter (1987) *How Experiments End*. Chicago: The University of Chicago Press.

Galison P. (1988) History, Philosophy, and the Central Metaphor. *Science in Context*, 2(1), pp.197-212.

Gingerich, Owen (1997) *The Eye of Heaven*. Berlin Springer.

Gingerich, Owen Jay (2004) *The Book Nobody Read*. New York: Walker.

Glazebrook Trish (2000) *Heidegger's Philosophy of Science*. Fordham University Press, New York, 278 p.

Goddu, Andrew (2010) *Copernicus and the Aristotelean Tradition: Education, Reading, and the philosophy in Copernicus's Path to Heliocentrism*. Leiden, Boston: Brill.

Goldberg, Stanley (1970) In defence of aether: the British response to Einstein's special theory of relativity, 1905-1911. *Historical Studies in the Physical Sciences,* vol. 2, pp. 89-125. Goldberg, Stanley (1976) Max Planck's philosophy of nature and his elaboration of the special theory of relativity. *Historical Studies in the Physical Sciences,* vol. 7, pp. 125-160.

Gould, A. (1987) “Classical derivation of black-hole entropy.” *The Physical Review* D35, no 2: 449-454.

Gravner J. and D. Griffeath (2009) Modeling Snow-Crystal Growth: A Three-Dimensional Mesoscopic Approach. *The Physical Review* E79: pp.1-18.

Hacking I. (1979) Michel Foucault’s immature science. *Nous* 13: 39-51.

Hacking, Ian (1987) Was There a Probabilistic Revolution, 1800-1930? – In: *Probabilistic Revolution,* vol.1: Ideas in History (Ed. by L. Kruger, L.J. Daston, and M. Heidelberger). Cambridge, Mass.: The MIT Press, pp.45-58.

Harman P.M. (2001) *The Natural Philosophy of J.C. Maxwell.* Cambridge University Press. - 228 p.

Hawking, S.W. (1975) “Particle Creation by Black Holes.” *Communications in Mathematical Physics* 43: 199-220.

Heidegger M. (1927) Sein und Zeit. *Jahrbuch für Philosophie und phänomenologische Forchung.* Band VIII. Halle: Max Nimeyer.438s.

Heidegger M. (1954) Die Frage nach der Technik. -In: *Die Künste im technischen Zeitalter.* München, S.70-108.

Heidegger M. (1954) Wissenschaft und Verständis.-In: *Vorträge und Aufsätze.* Pfullingen, Neske. S.45-70.

Heidegger Martin (1967) *What is a Thing?* Chicago: Henry Regnery.

Heidegger Martin (1971). *Was heist Denken?* Max Niemeyer Verlag Tubingen. – 175 Seiten.

Heidegger Martin ([1950], 1972) Die Zeit des Weltbildes, in: *Holzwege* (Frankfurt: V. Klostermann.

Heidegger, Martin ([1954], 1977) The Question Concerning Technology. – In: *The Question Concerning Technology and Other Issues,* translated by William Lovitt. New York: Harper & Row.

Heidegger, Martin, “Die Frage nach der Technik,” *Humanitas* 14 (1978): 17.

Heidegger, Martin (1995) *The Fundamental Concepts of Metaphysics: World, Finitude, Solitude*. Bloomington, IUP.

Heidegger, Martin ([1989], 1999). *Contributions to Philosophy (From Enowning)*. Bloomington: Indiana University Press.

Heidegger, Martin ([1927], 2001a). *Sein und Zeit.* Max Niemeyer.- 458 Seiten.

Heidegger, Martin (2001b). On the Essence of Truth. – In: *The Nature of Truth* (ed. by M. Lynch). The MIT Press. – pp. 295-316.

Heilbron, John L. and Thomas Kuhn (1969) The Genesis of the Bohr Atom. *Historical Studies in the Physical Sciences,* vol. 1, pp. 211-290.

Heilbron, John (2017) Where We Start? – In: Blum A., K. Gavroglu, C. Joas, J. Renn (eds.) *Shifting Paradigms. Thomas S. Kuhn and the History of Science*. Proceedings of the Conference "50 years Since Structure: Towards a History of the History of Science," held in Berlin in October 2012.Editor Open Access. Max Planck Institute for the History of Science, pp. 21-28.

Heisenberg, Werner and Pascual Jordan (1926) Andwendung der Quantenmechanik aufder Problem der anomalen Zeemaneffekte. *Zeitschrift fur Physik*, vol.37, 263-277.

Heisenberg, Werner (1959) Grundlegende Voraussetzungen in der Physik der Elementarteilchen.- In *Martin Heidegger zum Siebzigsten Geburtstag Festschrift*, ed. by Günther Neske. Pfullingen: Verlag Gunther Neske, pp. 291-97.

Heisenberg, Werner, (2000) *Physik und Philosophie*. Stuttgart: Hirzel.

Hertz H (1884) Ueber die Beziehungen zwischen den Maxwell'schen elektrodynamichen Grundgleichungen und den Grundgleichungen der gegnerischen Elektrodynamik. *Annalen der Physik und Chemie*, 23: 84-103.

Hirosige, Tetu (1976) "The ether problem, the mechanistic world view and the origins of the theory of relativity." *Historical Studies in the Physical Sciences*, 7: 3-82.

Holton G. (1968) Mach, Einstein and the Search for Reality. *Daedalus* 97: 636-673.

Holton, Gerald (2017) Steve's Question and Tom's last Lecture: A Personal Perspective. – In: Blum A., K. Gavroglu, C. Joas, J. Renn (eds.) *Shifting Paradigms. Thomas S. Kuhn and the History of Science.* Proceedings of the Conference "50 years Since Structure: Towards a History of the History of Science," held in Berlin in October 2012.Editor Open Access. Max Planck Institute for the History of Science, pp. 32-45.

Hon, Giora, and Bernard R. Goldstein (2012) Maxwell's contrived analogy: An early version of the methodology of modeling. *Studies in History and Philosophy of Modern Physics* 43: 236–257.

Hoyningen-Huene, Paul (1993) *Reconstructing Scientific Revolutions: Thomas S. Kuhn's Philosophy of Science*. Chicago: The University of Chicago Press.

Howard D. (1994) Einstein, Kant and the Origins of Logical Positivism. In: *Language, Logic and the Structure of Scientific Theories: The Carnap -Reichenbach Centennial*, ed. Wesley Salmon and Gedeon Wolters, 45-105. Pittsburgh: University of Pittsburgh Press.

Husserl, Edmund (1913) *Ideen zu einer reinen Phanomenologischen Philosophie* (Jahrbuch fur Philosophie and phanomenologische Forschung.Bd. I, Halle.

Husserl, Edmund. [1936], 1970) *The Crisis of European Sciences and Transcendental Phenomenology* (translated by David Carr). Northwestern University Press

Husserl Edmund (2007) *Die Krisis der Europaischen Wissenschaften und die Transzendentale Phanomenologie*. Meiner F. – 160 Seiten.

Husserl Edmund (2009) *Ideen zu einer reinen Phanomenologie und phanomenologischen Philosophie*. Meiner F. – 462 Seiten.

Ivins W.M. (1973) *On the Rationalization of Sight.* New York: Plenum Press.

Janssen M., Norton John D., Renn Jürgen, Sauer Tilman, Stachel John (2007) Introduction to volumes 1 and 2: the Zurich Notebook," in Boston Studies in the Philosophy of Science, vol. 250, Dordrecht: Springer, pp.7-20.

Janssen, Michel (2012) The twins and the bucket: How Einstein made gravity rather than motion relative in general relativity," *Studies in*

History and Philosophy of Science, Part B: Studies in History and Philosophy of Modern Physics, 43, 159-175.

Jeans, James (1905) On the Partition of Energy between Matter and Aether. *Philosophical Magazine* 10: 91-98.

Jones, Alexander (2005) Ptolemy's Mathematical Models and their Meaning. In *Mathematics and the Historian's Craft*, edited by M. Kinyon and G. van Brummelen. Springer, pp.23-42.

Kant I. ([1787]/1998) *The Critique of Pure Reason*. Repr. Translated by Paul Guyer and Allen W. Wood. Cambridge: Cambridge University Press.

Kant, Immanuel ([1783] /2002) Prolegomena to any future metaphysics that will be able to come forward as science. (Transl. by Gary Hatfield). – In: *Immanuel Kant. Theoretical Philosophy after 1781*. Cambridge University Press, 2002, pp.29-170.

Keller, Reiner (2011) The Sociology of Knowledge Approach to Discourse. *Hum. Stud*. 34:43-65.

Kellert, Stephen, Helen Longino and C. Kenneth Waters, eds. (2006) *Scientific Pluralism* (Minnesota Studies in the Philosophy of Science, vol. XIX), Minneapolis University of Minnesota Press.

Kelvin, Lord (1901) "19-th Century Clouds over the Dynamical Theory of Heat and Light." *Philosophical Magazine* 2:1-39.

Kepler, Johannes ([1609]. 2005) Astronomia Nova. In *Selections from Kepler's Astronomia Nova: A Science Classics Module for Humanities Studies* (translated by W.H. Donahue). Santa Fe, NM: Green Lion Press.

Kilmister C.V. (1994) *Eddington's search for a fundamental theory. A key to the universe*. Cambridge University Press.

Kitcher, Philip (1993) *The Advancement of Science: Science Without Legend, Objectivity Without Illusions*. Oxford: Oxford University Press.

Kochan, Jeff (2017) *Science as Social Existence. Heidegger and the SSK*.Open Book Publishers.

Kojeve Alexandre (1980). *Introduction to the Reading of Hegel: Lectures on the Phenomenology of Spirit.* Cornell University Press. – 287 p.

Kokowski, Michal (1996) Copernicus and the Hypothetico-Deductive Method of Correspondence Thinking. *Theoria et Historia Scientiarum*, vol. V: 7-103.

Kokowski, Michal (2012) Copernicus, Arabic Science, and the Scientific Revolution. In *Asia, Europe and the Emergence of Modern Science*, edited by A. Bala, 55-103.

Koyrè A. (1939) *Etudes galileennes*. Paris, Hermann.

Koyrè A. (1943) Galileo and Plato. *Journal of the History of Ideas*, 4:400-428.

Koyré, Alexander (1957) *From the Closed World to the Infinite Universe.* Baltimore: John Hopkins Press

Koyrè A. (1961) *La revolution astronomique: Copernic, Kepler, Borelli.* Paris: Hermann.

Koyrè A. (1973) *The Astronomical Revolution. Copernicus-Kepler-Borelli.* N.Y. Dover.

Kragh, Helge (1981) The Genesis of Dirac's Relativistic Theory of Electrons. *Archive for History of Exact Sciences*, vol. 24, 31-67.

Kragh, Helge (1990) *Dirac. A Scientific Biography*. New York: Cornell University.

Kragh, Helge S. and James M. Overduin (2014) *The Weight of the Vacuum: A Scientific History of Dark Energy.* Heidelberg: Springer.

Kuhn, T.S. (1957) *The Copernican Revolution: Planetary Astronomy in the Development of Western Thought.* Cambridge: Harvard University Press.

Kuhn, T.S. (1963.) The Function of Dogma in Scientific Research. In *Scientific Change*, edited by A.C. Crombie. London: Heinemann.

Kuhn, Thomas S. (1962], 1970) *The Structure of Scientific Revolutions*. Second Edition, Enlarged. The University of Chicago Press.

Kuhn T.S. (1976) Mathematical vs. Experimental Traditions in the Development of Physical Science. *The Journal of Interdisciplinary History*, 7: 1-31.

Kuhn T.S. (1977) Objectivity, Value Judgement and Theory Choice. - In: *The Essential Tension.* University of Chicago Press, pp. 320 – 339

Kuhn T.S. (1978) *Black-Body Theory and Quantum Discontinuity, 1894-1912*. Oxford and New York: Oxford University Press.

Kuhn T.S. (1993) Afterwords. – In: *World Changes: Thomas Kuhn and the Nature of Science*. Cambridge: MIT Press.

Kuhn, Thomas S. (2000) *The Road Since Structure*. Chicago: The University of Chicago Press

Lakatos, Imre (1970) Falsification and the Methodology of Scientific Research Programmes. In *Criticism and the Growth of Knowledge*, edited by I. Lakatos and A. Musgrave, 91-195. Cambridge: Cambridge University Press.

Lakatos, Imre and Elie Zahar (1974) Why did Copernicus's Research Program Supersede Ptolemy's? In *The Copernican Achievement*. University of California, Los Angeles, ch. X, 168-192.

Lakatos, Imre ([1970], 1978) *The Methodology of Scientific Research Programmes*. Philosophical Papers, volume 1. Edited by J. Worral & G. Currie, Cambridge University Press, 1978. - 254p.

Landau, Lev (1958) The Quantum Theory from Max Planck up to now. In *Max Planck.1858-1958*, 93-98. Moscow: USSR Academy of Science Publishing House (in Russian).

Landau, Lev D. and Lifshitz, Eugene M. (1973) *Theory of Fields*. L.: Pergamon Press.

Latour, B. (1983) Comment redistribuer le grand partage? *Revue Internationale de Synthese* 104 (110): 202-236.

Latour, B., and S. Woolgar ([1979], 1986) *Laboratory Life: The Social Construction of Scientific Facts*. London: Sage, 2nd edition: Princeton: Princeton University Press.

Latour B. (1990) Drawing things together. In *Representation in Scientific Practice*, edited by M. Lynch and S. Woolgar. Massachusetts: The MIT Press, 19-68.

Laudan L. (1981) The pseudo-science of science? *Philosophy of the Social Sciences* 11(2): 173-198.

Lial, Margaret L., Miller Charles David, Hornsby John E. (1995) *Beginning Algebra*. Harper Collins College Publishers.

Linton, Cristopher M. (2010) *From Eudoxes to Einstein: a history of mathematical astronomy.* Cambridge University Press.

Longino, Helen E. (1990) *Science as Social Knowledge: Values and Objectivity in Scientific Inquiry.* Princeton: Princeton University Press.

Lorentz, H.A. (1909) *The theory of electrons and its applications to the phenomena of light and radiant heat.* New York: G.F. Stechart & Co.

Lynch M. and Woolgar S. (1990) Introduction. Sociological orientations to representational practice in science. In *Representation in Scientific Practice*, edited by Michael Lynch and Steve Woolgar. Cambridge, Mass.: The MIT Press, 1-18.

Mach E. (1942) *The Science of Mechanics: A Critical and Historical Account of Its Development.* London: Open Court Publishing Company.

Mach E ([1897], 1984) *The Analysis of Sensations and the Relation of the Physical to the Psychical.* Translated by C.M. Williams. La Salle: Open Court.

Mach E. ([1893] / 1999) *The Science of Mechanics: A Critical and Historical Account of its Developments.* (Translated by T. J. Mc Cormmach). La Salle: Open Court.

Mahon, Basil (2003) *The Man Who Changed Everything. The Life of James Clerk Maxwell.* John Wiley. – 254p.

Maxwell J.C. ([1856], 1890) On Faraday's Lines of Force. *The Transactions of the Cambridge Philosophical Society*, vol. X, part 1. Reprinted in "The Scientific Papers of James Clerk Maxwell," 1890, vol. 1, pp. 155-229.

Maxwell J.C. ([1861], 1890) On Physical Lines of Force. *Philosophical Magazine*, 1861, vol. XXI. Reprinted in "The Scientific Papers of James Clerk Maxwell," 1890, vol.1, pp.451-513.

Maxwell J.C. ([1864], 1890) A Dynamical Theory of Electromagnetic Field, 1864. Reprinted in "The Scientific Papers of James Clerk Maxwell," 1890, vol. 1, pp.526-597.

Maxwell J.C. ([1873b], 1890) On Action at a Distance. Reprinted in "*The Scientific Papers of James Clerk Maxwell*," 1890, vol. 1, pp.315-320.

Maxwell J.C. ([1877], 1890) Hermann Ludwig Ferdinand Helmholtz. *Nature*, vol. XV. Reprinted in "The Scientific Papers of James Clerk Maxwell," 1890, vol.2, pp.592-598.

Maxwell J.C. ([1873a], 1998) *A Treatise on Electricity and Magnetism*. Clarendon Press, 1998. – 532p.

Mehra, Jagdish and Helmut Rechenberg (2000) *The Historical Development of Quantum Theory*. Vol.6. The Completion of Quantum Mechanics 1926-1941. Part I. The Probability Interpretation and the Statistical Transformation Theory. New York: Springer-Verlag.

Mertz, John Theodore (1964) *A History of European Thought in the Nineteenth Century*, 4 vols. Edinburgh: William Blackwood and Sons, 1903-1912. Vols 1 and 2 reprinted as "A History of European Scientific Thought in the Nineteenth Century." NY.

Meyerson, Emile (1912) *La deduction relativist*. Paris: Pavot.

Miller A. (1981) *Albert Einstein's Special Theory of Relativity*. London: Addison-Wesley.

Morton Wise, N. (2017) Goethe Was Right: The History of Science is Science Itself. – In: Blum A., K. Gavroglu, C. Joas, J. Renn (eds.) *Shifting Paradigms. Thomas S. Kuhn and the History of Science.* Proceedings of the Conference "50 years Since Structure: Towards a History of the History of Science," held in Berlin in October 2012.Editor Open Access. Max Planck Institute for the History of Science, pp. 347-360.

Morrison, Margaret (2000) *Unifying Scientific Theories: Physical Concepts and Mathematical Structures*. Cambridge University Press. - 270p.

Mulkay, Michael (2015) *Science and the Sociology of Knowledge*. London: Routledge.

Murschel, A. (1985) "The structure and function of Ptolemy's physical hypotheses of planetary motion." *Journal for the History of Astronomy*, XXVI: 33-61.

Neugebauer, Otto (1975) *A History of Ancient Mathematical Astronomy*. 3 vols. Berlin: Springer. Newton, Isaac ([1687], 1846) *The Mathematical Principles of Natural Philosophy* (translated into

English by Andrew Motte, with preface of Mr. Roger Cotes). New York: Daniel Ades.

Nickles T. (1986) Remarks on the Use of History as Evidence. *Synthese* 69(2), pp. 253-266.

Niven W.D. (1890) A Preface to "The Scientific papers of J.C. Maxwell." Dover publications, NY, 1890, vol.1, pp. XI-XIII.

Nordström, Gunnar (1912) "Relativitätsprinzip und Gravitation," *Physikalische Zeitschrift*, 13, 1126-1129. Translated in Jürgen Renn, ed., The Genesis of General Relativity. Vol.3.Gravitation in the Twilight of Classical Physics: Between Mechanics, Field Theory and Astronomy. Dordrecht: Springer, 2007.

Nordström, Gunnar (1913a) Trage und Schwere Masse in der Relativitätsmechanik, *Annalen der Physik*, 40, 856-878.

Nordström, Gunnar (1913b) Zur Theorie der Gravitation vom Standpukt des Relativitätsprinzip, *Annalen der Physik,* 42, 533-534.

Norton, John D. (1986) What was Einstein's Principle of Equivalence Principle? – In: *Einstein's Studies*, vol.1

Norton, John D. (1992) "Einstein, Nordström and the Early Demise of Scalar, Lorentz Covariant Theories of Gravitation," *Archive for the History of Exact Sciences*, 45: 17-94. Reprinted in: Jürgen Renn, ed., *The Genesis of General Relativity*. Vol.3.Gravitation in the Twilight of Classical Physics: Between Mechanics, Field Theory and Astronomy. Dordrecht: Springer, pp. 413-488.

Norton Wise M. (2017) Goethe Was Right: The History of Science is Science Itself' – In: Blum A., K. Gavroglu, C. Joas, J. Renn (eds.) *Shifting Paradigms. Thomas S. Kuhn and the History of Science.* Proceedings of the Conference "50 years Since Structure: Towards a History of the History of Science," held in Berlin in October 2012.Editor Open Access. Max Planck Institute for the History of Science, pp. 347-360.

Nugayev R. M. (1985a) A Study of Theory Unification // *The British Journal for the Philosophy of Science*, vol. 36, pp. 159 – 173.

Nugayev Rinat, M. (1985b) The History of Quantum Theory as a Decisive Argument Favoring Einstein over Lorentz. *Philosophy of Science*, vol. 52, pp. 44 – 63.

Nugayev, Rinat M. (1999) *Reconstruction of Mature Theory Change: A Theory-Change Model*. Frankfurt am Main: Peter Lang.

Nugayev, Rinat M. (2012) *Copernican Revolution: Unification of Mundane Physics with Mathematics of the Skies*. Kazan, Innovative Technologies Publishing House, 2012. – 302 p. (in Russian).

Nugayev, Rinat M. (2013) "The Ptolemy-Copernicus Transition: an Intertheoretical Context." *Almagest*, vol. IV (1), pp.96-119.

Nugayev, Rinat M. (2015) Communicative Rationality of the Maxwellian Revolution. *Foundations of Science*, vol. 20, issue 4, 447-478. DOI: 10.1007/s10699-015-9417-7

Nugayev, Rinat M. (2018) *Einstein's Revolution: A Study in Theory Unification*. Bentham Science Publishers. Sharjah: UAE-USA.

Nugayev, Rinat M. (2019) Einstein's 1905 Annus Mirabilis: Reconciliation of the Basic Research Traditions of Classical Physics. *Axiomathes,* vol.29, pp.207-235.

Nugayev, Rinat M. (2020) *The Planck-Einstein Breakthrough: Reconciliation of the Pivotal Research Programs of the Classical Physics*. Montreal: Minkowski Institute Press.

Nugayev, Rinat M. (2022) Dirac's Refined Unification of Quantum Mechanics and Special Relativity: an Intertheoretic Context. *Teorie Vedy* (Theory of Science), vol. 44, no 1, pp. 1-22.

Nugayev, Rinat M. (2023) The Copernican Long-Awaited Breakthrough: Reconciliation of the Blunt Earth Physics and the Immaculate Mathematics of the Skies. *Cosmos & History: The Journal of Natural and Social Philosophy*, vol.19, no 2, pp. 255-278.

Olson, Richard (1975) *Scottish Philosophy and British Physics, 1750-1880: A Study in the Foundations of the Victorian Scientific Style*. Princeton, NJ, Princeton University Press, 1975. – 349p.

Omodeo, Pietro Daniel (2017) Kuhn's Paradigm of Paradigms: Historical and Epistemological Coordinates of the Copernican Revolution. – In: Blum A., K. Gavroglu, C. Joas, J. Renn (eds.) *Shifting Paradigms. Thomas S. Kuhn and the History of Science*. Proceedings of the Conference "50 years Since Structure: Towards a History of the History of Science," held in Berlin in October 2012.Editor Open Access. Max Planck Institute for the History of Science, pp. 71-104.

Pais A. (1979) Einstein and the Quantum Theory. *Rev Mod Phys*, 51: 863-914.

Pais A. (1982) *Subtle is the Lord. The Science and Life of Albert Einstein.* Oxford: Oxford University Press.

Pauli, Wolfgang ([1921], 1981) *Theory of Relativity.* N.Y.: Dover Publications.

Pauli, Wolfgang (1926) Über das Wasserstoffspektrum vom Standpunkt der neuen Quantenmechanik. *Zeitschrift fur Physik*, vol. 36, 336-363.

Payne W. Russ (2015) *An Introduction to Philosophy*. Bellevue College.

Petkov, Vesselin (2014) *Physics as Spacetime Geometry*. Published as Chapter 8 in Springer Handbook of Spacetime. Springer: Heidelberg, pp. 141-163.

Perlmutter, S. et al. (1999) "Measurements of ω and λ from 41 high-redshift supernovae." *The Astrophysical Journal* 517:565-586.

Pickering, Andrew (1985) *Constructing Quarks. A Sociological History of Particle Physics.* Chicago: The University of Chicago Press.

Planck, Max (1897a): Uber irreversible Strahlungsvorgange. Erste Mitteilung. *Berl. Ber.*, pp.57-68.

Planck, Max (1897b): Uber irreversible Strahlungsvorgange. Zweite Mitteilung. *Berl. Ber.*, pp.715-717.

Planck, Max (1897c): Uber irreversible Strahlungsvorgange. Dritte Mitteilung. *Berl. Ber.*, pp.1122-1145.

Planck, Max (1898): Uber irreversible Strahlungsvorgange. Vierte Mitteilung. *Berl. Ber.*, pp.57-68.

Planck, Max (1899): Uber irreversible Strahlungsvorgange. Funfte Mitteilung (Schluss). *Berl. Ber.*, pp.440-480.

Planck, Max (1900a): Uber irreversible Strahlungsvorgange. *Annalen der Physik*, vol.1, pp.69-122.

Planck, Max (1900b): Zur Theorie des Gesetzes der Energieverteilung im Normalspectrum. *Verh. D. D. Phys. Ges.*, vol.2, pp. 237-245. Translated in Ter Haar, 1967.

Planck, Max (1906): Vorlesungen uber die Theorie der Warmestrahlung. First edition. Leipzig: J.A. Barth.

Planck, Max (1910): *Acht Vorlesungen uber theoretische Physik*. Leipzig: Hirzel. Originally delivered at Columbia university, 1909, and translated as "Eight Lectures on Theoretical Physics" by A.P. Wills, New York, Columbia University Press, 1915.

Planck, M. (1911) "Eine Neue Strahlungshypothes." *Verh. Dtsch. Phys. Ges.* 13:138-148.

Planck, Max (1922) Die Einheit des physikalischen Weltbildes. Physikalische Rundblicke. - Leipzig

Planck, Max (1948) *Wissenchaftlische Selbstbiographie*. Leipzig: Barth.

Planck, Max ([1908]'1960) *A Survey of Physical Theory*. Translated by R. Jones and D.H. Williams. NY: Dover Publications.

Planck M. (2001) *Vorträge, Reden, Erinnerungen* /hrsg. Von Hans Roos; Armin Hermann, Berlin; Heidelberg:Springer Verlag.

Plato. *Timaeus* ([386 BC], 2000) (translated by D.J. Zeyl). Indianapolis and Cambridge, Massachusetts: Hackett Publishing Co.

Podgoretzky M.I., Smorodinsky Ja. A. (1980) On Axiomatic |Structure of Physical Theories. – In *Physical Theory*. Moscow: Nauka, pp. 53 – 61 (in Russian).

Poincaré, H. ([1902], 1905) Science *and Hypothesis.* Repr. London and Newcastle-on-Tyne: The Walter Scott Publishing Co.

Poincaré, Henri (1906) Science *et Hypothese.* Paris: Flammarion.

Popper, Karl ([1935], 2002) *The Logic of Scientific Discovery*. London and New York: Routledge.

Power Michael. 2011.Foucault and Sociology. *Annual Review of Sociology*, 37: 35-56.

Ptolemy, Claudius (1998) *The Almagest* (translated by G.J. Toomer with a foreword by Owen Gingerich). Princeton University Press.

Putnam, Hilary (2001) "Two Philosophical Perspectives." In *The Nature of Truth. Classic and Contemporary Perspectives*, edited by M. Lynch, 251-258. Cambridge, Massachusetts: The MIT Press.

Quine, W.V. ([1951], 2000) "Two Dogmas of Empiricism." In *Contemporary Analytic and Linguistic Philosophies*, edited by E.D. Klemke, 331-349. New York: Prometheus Books

Renn, Jurgen (1993) Einstein as a Disciple of Galileo: a Comparative Study of Concept Development in Physics. *Science in Context* 6, 1, pp. 311-341.

Renn, Jurgen (1996) *Historical Epistemology and the Advancement of Science*. Max Planck Institute for the History of Science. Preprint 36.

Renn, Jurgen (2004) The Relativity Revolution from the Perspective of Historical Epistemology. *Isis*, vol. 95, pp.640-648.

Renn, Jürgen (2007a) "Classical Physics in Disarray. The Emergence of the Riddle of Gravitation," in Jürgen' Renn, ed., 2007. *The Genesis of General Relativity*. vols. 1-2, Dordrecht: Springer, pp.21-84.

Renn, Jürgen and Sauer, Tilman (2007) "Pathways out of Classical Physics: Einstein's Double Strategy in his Search for the Gravitational Field Equations," in Jürgen Renn, ed., *The Genesis of General Relativity*. Vol. 1. Dordrecht: Springer, pp. 113-312.

Renn, Jurgen and Schulmann Robert (1992) "Introduction." In: Albert Einstein/ Mileva Maric *The Love Letters*. Ed. and with an introduction by Jurgen Renn and Robert Sculmann. Princeton NJ: Princeton University Press

Rheinberger, Hans-Jorg (2009) *Towards a History of Epistemic Things: Synthesizing Proteins in the Test Tube*. Stanford: CA: Stanford University Press.

Rouse, Joseph (1987) *Knowledge and Power: Toward a Political Philosophy of Science*. Ithaca: Cornell University Press.

Rouse, Joseph (2002) Kuhn: Philosophy of Scientific Practice. In *Thomas Kuhn*, edited by Thomas Nickles, 100-121. Cambridge: Cambridge University Press.

Rouse, Joseph (2013) Recovering Thomas Kuhn, *Topoi*, 32 (1), pp.59-64.

Rouse, Joseph (2014) Scientific Practice and the Scientific Image. -In: Lene Soler, Sjoert Zwart, Michael Lynch, and Vincent Israel-Jost (eds.) *Science after the Practice Turn in the Philosophy, History, and Social Studies of Science*. Chapter 9. L.: Routledge.

Rudwick, M.J.S. (1985) *The Great Devonian Controversy: The Shaping of Scientific knowledge among Gentlemanly Specialists.* Chicago: The University of Chicago Press.

Rudwick, Martin J.S. (2017) Constructive Controversy and the Growth of Knowledge. – In: Blum A., K. Gavroglu, C. Joas, J. Renn (eds.) *Shifting Paradigms. Thomas S. Kuhn and the History of Science.* Proceedings of the Conference "50 years Since Structure: Towards a History of the History of Science," held in Berlin in October 2012.Editor Open Access. Max Planck Institute for the History of Science, pp. 163-180.

Ryckman, Thomas (2005) *The Reign of Relativity. Philosophy in Physics 1915-1925.* - Oxford: Oxford University Press.

Rynasiewicz R. (2000) The Construction of the Special Theory: Some Queries and Considerations. In *Einstein. The Formative Years, 1879-1909*, ed. Don Howard & John Stachel, 159-201, Boston: Birkhauser.

Seelig, Carl (1957) *Helle Zeiten – Dunkle Zeiten: In memoriam Albert Einstein.* (Zurich /Stuttgart/ Wien: Europa Verlag.

Seelig, Carl (1960) *Albert Einstein. Leben und Werk Eines Genies Unserer Zeit.* Europa Verlag.

Schrödinger, Erwin (1926) Quantisierung als Eigenwertproblem. *Annalen der Physik*, 79, 361-376.

Schweber, Silvan S. (2017) On Kuhnian and Hacking-Type Revolutions. – In: Blum A., K. Gavroglu, C. Joas, J. Renn (eds.) *Shifting Paradigms. Thomas S. Kuhn and the History of Science.* Proceedings of the Conference "50 years Since Structure: Towards a History of the History of Science," held in Berlin in October 2012.Editor Open Access. Max Planck Institute for the History of Science, pp. 337-346.

Shapin S. and S. Schaffer (1985) *Leviathan and the Air-Pump: Hobbes, Boyle, and the Experimental Life.* Princeton: Princeton University Press.

Shapin S. (1996) *The Scientific Revolution.* Chicago: the University of Chicago Press.

Shapiro I.S. (1973) On the History of the Discovery of the Maxwell equations. *Soviet Physics Uspekhi,* vol.15, number 5, pp. 651-659.

Shea William (2017) Thomas Kuhn and the Dialogue Between Historians and Philosophers of Science. – In: Blum A., K. Gavroglu, C. Joas, J. Renn (eds.) (2017) *Shifting Paradigms. Thomas S. Kuhn and the History of Science.* Proceedings of the Conference "50 years Since Structure: Towards a History of the History of Science," held in Berlin in October 2012.Editor Open Access. Max Planck Institute for the History of Science, pp.163-180

Siegel, Daniel M. (1991) *Innovation in Maxwell's electromagnetic theory: molecular vortices, displacement current, and light.* Cambridge University Press. - 229 p.

Sigurdson Skuli (2017) The Nature of Scientific Knowledge. An Interview with Thomas S. Kuhn. – In: A. Blum, K. Gavroglu, C. Joas, J. Renn (eds.) *Shifting Paradigms. Thomas S. Kuhn and the History of Science.* Editor Open Access. Max Planck Institute for the History of Science.

Speziali P. (ed.) (1972) *Albert Einstein-Michele Besso: Correspondence.* Paris: Hermann. Stachel, John (1989) The Rigidly Rotating Disk as the 'Missing Link' in the History of General Relativity. In D. Howard and J. Stachel (eds.) *Einstein and the History of General Relativity.* (Einstein Studies, Vol.1). Boston, Basel and Berlin: Birkhauser, pp. 48-62.

Stachel, John (2017) History of Science: The French Connection. – In: Blum A., K. Gavroglu, C. Joas, J. Renn (eds.) (2017) *Shifting Paradigms. Thomas S. Kuhn and the History of Science.* Proceedings of the Conference "50 years Since Structure: Towards a History of the History of Science," held in Berlin in October 2012. Editor Open Access. Max Planck Institute for the History of Science, pp.301-328.

Stepin, Vyacheslav (2006) *Theoretical Knowledge*. Springer. Volume 326 of Synthese Library.

Suppe, Frederick ([1974], (1977) *The Structure of Scientific Theories.* The University of Illinois Press, 1977.

Swerdlow, Noel M. and Otto Neugebauer (1984) Mathematical Astronomy in Copernicus's "De Revolutionibus. *Studies in the History of Mathematics and Physical Science*, 10, 2 parts. N.Y.: Springer-Verlag.

Ter Haar D. (1967) *The Old Quantum Theory*. Oxford: Pergamon Press.

Tisza L. (1963) The Conceptual Structure of Physics // *Reviews of Modern Physics*, vol. 35, pp.151-185.

Unruh, W.G. (1980) Comment on Hawking's paper. In: *Some Strangeness in the Proportion*, edited by Harry Wolf, 153-155. N.Y.: Addison-Wesley.

Valente Mario (2020) The Dirac equation and its interpretations //URL: http: www.researchgate.net/publication/340664785, (date of application 03.03.2020)

Van Dongen J (2010) *Einstein's Unification*. Cambridge: Cambridge University Press.

Vizgin, V.P. (1981) *Relativistic theory of gravity: genesis and development.1900-1915.*Moscow: Nauka (in Russian).

Von Weizsäcker, Carl Friedrich ([1949], 1990) Beziehungen der theoretischen Physik zum denken Heideggers, in: Zum Weltbild der Physik, thirteenth edition, by Carl Friedrich von Weizsäcker. Stuttgard: S.Hirzel Verlag, pp.243-45.

Wartofsky, Marx W. (1979) *Models. Representation and the Scientific Understanding*. Springer: Dordrecht.

Weber, Max (1964) *Gesammelte Aufsätze zur Wissenschaftslehre*. Tübingen: Mohr.

Westman, Robert S. 1994. Two Cultures or One? A Second Look at Kuhn's "The Copernican Revolution." *Isis*, 85:79-11.

Weyl, Hermann ([1919], 1922) *Raum, Zeit, Materie*. (Berlin: Springer, 1919). Translated into English by Henry L. Brose as "Space-Time-Matter," L.: Methuen & Co, Ltd., 1922.

Wien W. '(1893) Eine neue Beziehung der Strahlung Schwarzer Köpperzumzweiten Hauptsatz der Wärmetheorie. *Annalen der Physik und Chemie* (Neue Folge), Bd. 49 (285), S.633-641.

Wien W. (1898) Ueber die Fragen, welche die translatorische Bewegung des Lichtathers betreffen. *Annalen der Physik und Chemie* Bd. 65 (3), S. i-xvii.

Whittaker Edmund Taylor (1910) *A history of the theories of aether and electricity: from the age of Descartes to the close of the nineteenth century*. L., N.Y., Longmans, Green and Co. – 470 p.

Wigner E. P. (1957) Relativistic invariance and quantum phenomena. *Reviews of Modern Physics*, vol.29, pp.255-268.

Woolgar S. (1981) Interests and explanation in the social study of science. *Social Studies of Science* 11: 365-394.

Woolgar S. (1983) Irony in the social study of science. In *Science observed: Perspectives on the social study of science*, edited by K. Knorr- Cetina and M. Mulkay. London and Beverly Hills: Sage

Worrall, John (2000). *Lakatos and After*. LSE Centre for the Philosophy of the Natural and Social Sciences.

Worrall, John (2020) Structural Realism: The Only Defensible Realist Game in Town? – In: J. Wenceslao (ed.) *New Approaches to Scientific Realism*, De Gruyter, pp. 169-205.

Wrathall, Mark (2011) A. *Heidegger and Unconcealment*. Cambridge University Press. -264 p.

Yang, C.N., and Robert Mills (1954) Conservation of Isotopic Spin and Isotopic Gauge Invariance. *The Physical Review*: 96-191.

Zahar, E. (1973) Did Einstein's programme supersede Lorentz's? The *British Journal for the Philosophy of Science*, 24: 95-123, 226-262.

Zahar, Elie (1989) *Einstein's Revolution: A Study in Heuristic*. La Salle: Open Court. Zel'dovich Ja. B., Novikov I.D. (1973) *Stars and Relativity*. L: Pergamon Press.

Zizek, Slavoj. 2009. *The Sublime Object of Ideology*. (The Essential Zizek). L.: Verso.

www.ingramcontent.com/pod-product-compliance
Lightning Source LLC
LaVergne TN
LVHW010718210926
848485LV00008B/1440/J